IBM's
SHADOW FORCE

Dust jacket art designed by Robert Mesa
Dust jacket layout by ThomasMax (Lee Clevenger & R. Preston Ward)
Photo of Mr. Robinson by Clark B. Savage
Edited by Lee Clevenger

Author's website: www.WLRobinson.com

First printing, April, 2008

Library Of Congress Cataloging-in-Publication Data:

Robinson, William Louis, 1928-
    IBM's *Shadow Force* : *the Untold Story of Federal Systems, the Secretive Giant That Safeguarded America* / William Louis Robinson.
    p.   cm.
    Includes bibliographical references and index.
    LCCN 2008903104
    ISBN-13: 978-0-9799950-3-3
    ISBN-10: 0-9799950-3-5

1. International Business Machines Corporation. Federal Systems Division--History.
2. Computer industry--United States--History.
I. Title.

HD9696.2.U64I256 2008   338.4'70040973   QBI08-600137

Published by:

tm

ThomasMax Publishing
P.O. Box 250054
Atlanta, GA 30325
404-794-6588
www.thomasmax.com

IBM's
SHADOW FORCE

The Untold Story of Federal Systems
The Secretive Giant That Safeguarded America

William Louis Robinson

ThomasMax

Your Publisher
For The 21st Century

ACKNOWLEDGEMENTS

While it is not possible to acknowledge all the assistance I received in the preparation of this book, I want to thank the following for their kind help, at the same time absolving them from any of my errors: Mr. Robert Campenni, Mr.Buzz Bernard, Mr. George Liptak, Mr. Steve Jackson, Mr. Fred Lippucci, and special thanks to my brother and his wife, Jack and Mary Robinson, and all the ex-IBMer's at the OWL group. And most especially, my wife, Grace, for all her help and patience.

This book is for Grace

Table Of Contents

*He who fashions the battle sword is as much a warrior
as the soldier who wields it*

PREFACE

During the past twenty-five years, at least nineteen books containing millions of words have been written about IBM. As I write this, the nineteen I located are stacked before me. I have read every word. In all of these thousands of pages, the words "Federal Systems Division," "Federal Systems Company," or "FSD" appear only eight times. In fact, descriptions of IBM's decades-long military/government work amount to only a dozen or so paragraphs — most covering FSD's demise. Why? Was it by design? Why did not one author care or dare to write details of its existence? The Manhattan Project had more print copy than this brilliant group of IBMers and their technological service to the nation. Was the U.S. government somehow behind this lack of public knowledge or was it something else? The eleven-page *History of IBM* on the Internet's Wikipedia encyclopedia has but a simple two-sentence paragraph on FSD! Why?

Having spent 17 years with IBM, I decided to find out. What I learned is the subject of this book. I interviewed many former and retired IBMers. Amazingly, most could shed little light on one of IBM's most ingenious and advanced operating organizations. FSD was not some small group laboring in an obscure lab in the boondocks. This division was composed of thousands of scientists and engineers located throughout the world performing technological breakthroughs in computer communication, space flight systems, satellite engineering, submarine acoustics, ballistic missile surveillance and defense command and control systems — including ultra-secret computing for DOD, CIA, NSA and FBI (NSA alone had ten acres of computers).

FSD had one of the largest concentrations of degreed

engineering talent in the corporation. IBM skirted bankruptcy in the early 1990s, saved by a strong new CEO, Louis V. Gerstner, Jr., by changing its strategic business model and by eliminating large portions of its workforce. IBM had lost $2.8 billion in 1991, $5 billion in 1992, and a staggering $8.1 billion in 1993.

After just several months of study, and pressured by a severe cash crisis, the Federal Systems organization was put on the block and sold to Loral Corporation for $1.5 billion. This decision was consummated just after FSD had received a one-half billion dollar Army contract for modernizing computers at 128 bases; it also became prime supplier of the Navy's next advanced AN/AYK-14 airborne computer, and won an IRS $1.3 billion, 15-year cost-plus award contract with an IRS down payment to Loral of $66 million.

CFO Jerome York, a "cost-cutter" specialist, had been newly hired by Lou Gerstner to find money; his orders were to sell whatever assets he could find to right the ship. So FSD was jettisoned without, in this writer's opinion, a full comprehension of the experienced and specialized talent being cast adrift. This know-how and experience would have been of enormous value to the IBM Corporation as it began its strategic makeover, repositioning resources to address global services fee-based consulting. One former FSDer, Dennie M. Welsh, a highly talented senior manager from FSD's NASA Apollo space contract, persisted and finally convinced Lou Gerstner to make the services business a major focus. The new CEO did . . . but long after FSD was gone! Today, half of IBM's business is derived from consulting services.

For well over three decades, Federal Systems people had been out front, performing advanced-systems information-handling contracts and consulting services to federal, state and local governments, to U.S. industry, and to foreign businesses and governments — all without public recognition. Little is known of this historic work done by so many talented FSD people. This book spotlights much of these endeavors. It proclaims FSD's quiet legacy to the IBM of today.

Ironically, two years after buying FSD, Loral divested its entire defense electronics and system integration business (including IBM's Federal Systems operations) to Lockheed Martin for $9.1 billion. Lockheed Martin is now the world's #1 military contractor with sales of $39.6 billion and a backlog of $75.9 billion. Only Lockheed Martin knows how many FSDer's contributed to this success. I know of one who made it up to senior vice president and chief financial officer, headquartered in FSD's former Owego, New York facility, now named Lockheed Systems Integration enterprise.

A major reason IBM recovered from the disasters of the early 1990s was its decision to make paid consultant services a major focus of its business — pioneered by the thousands of contractual projects, ideas, skills, and talents of unheralded Federal Systems people.

Who then were these FSD people? What was it they did? How did they do it? Who were the customers that benefited and how much in the way of standard computer product sales did FSD work contribute to its parent IBM? What real service to the nation did this remarkable group of people perform with little recognition?

This book is not a history of the IBM corporation. Nor is it a history of computer development. It does not delve into the reasons for IBM's downfall nor its remarkable phoenix-like rise under Lou Gerstner's leadership. It is the story of a group of people doing extraordinary work to protect our nation. Until now no author has written a comprehensive history of the role FSD played in our nation's security, space endeavors and international business.

Anyone who has worked for IBM knows there is a code of secrecy surrounding its business operations. The company dislikes publicity. Much of this has had to do with the highly competitive nature of the computer industry. Also, most government work, when not specifically classified, tends to be compartmentalized by defense contractors performing technical work. Besides, FSD employees were usually too busy under tight schedules to worry about what was going on "next door." Much, however, was classified and the IBM corporation itself had a Top Secret clearance granted by the Air Force in 1955.

These are some of the reasons that prior researchers and authors of volumes about IBM have never discussed in detail the Federal Systems Division and its highly talented work force. FSD, from its beginnings, was an organization virtually unknown to the public. It was the precursor to today's highly successful fee-based IBM Global Consulting services business. The time has come to remove the shroud of obscurity and proclaim Big Blue's quiet legacy. FSD: the IBM nobody really knew.

Almost a decade after Federal Systems was sold, an IBM spokesman told reporters that Federal Systems was sold because it had strayed too far from IBM's core information technology business even though FSD had been consistently providing widening profit margins from just that one segment alone: systems integration and services within the federal marketplace. The year FSD was sold, the government had budgeted over $25 billion for information technology services. Chapter eight of this book provides compelling evidence that FSD was the internal IBM group that IBM domestic divisions and its international arm, World Trade companies, sought help in solving complex technical problems in areas outside of federal procurement. All that technical expertise now benefits Lockheed Martin that today is the world's leading defense contractor in aeronautics, homeland security systems, space systems, and postal automation. Although a large percentage of FSD people are now retired, many are still hard at work for a new boss and for our nation's defense.

I came out of the Navy a senior non-commissioned officer in 1954 full of enthusiasm and eager to tackle the commercial world. As a veteran of both World War II and the Korean War, I had spent more than three years on a heavy cruiser in the Atlantic and Mediterranean, was carefully vetted for a two-year duty tour at the Military Mission in Ankara, Turkey (established under the Truman Doctrine to save Greece and Turkey from Soviet influence), and a final tour at the Pentagon as one of four aides to Admiral Donald B. Duncan, the Vice Chief of Naval Operations.

The Korean War was raging and battle reports from General Ridgeway crossed my desk daily from the Joint Chiefs of Staff Directorate. Dwight D. Eisenhower was President and a direct White House telephone line connected our Pentagon CNO suite.

I mention this to provide some credibility to the story I am going to relate about IBM. I interacted with admirals and senior government officials throughout my eight-year naval career, and while non-commissioned, I acted on behalf of and with great authority for them. The same was true during my seventeen-year career with IBM, primarily with FSD, the government sales division, although I was not a senior executive.

From January 1955, when I joined IBM in the Washington, D.C., office, I worked my way up through customer engineering support and computer orders and movements for all U.S. Army Post Camps and Stations worldwide, to personal aide to the IBM Vice President/Director of Federal Government Operations, Mr. M.B. Smith, to Manager of FSD's Technical Proposal Department and eventually to Manager of Federal Market Analysis for the organization. My final two years were spent managing the administrative aspects of FSD's technical support to IBM's international World Trade Corporation countries throughout the globe . . . from Europe to the Far East.

I became an expert in the drafting of R&D technical proposals to federal government agencies and other aerospace defense contractors. I wrote three manuals on the art of selling proposals. One was internally and the other two commercially published. I spoke at industry gatherings and IBM sales schools. I prepared analyses of civil government and military programs, spent time on Capitol Hill gathering and analyzing budgetary information and traveled internationally, presenting FSD's capabilities to IBM's World Trade companies in Canada, Europe, and South America.

Because of this broad experience I realized that few people in or outside of FSD had such a grand view of the capabilities, talents and skills inherent in an organization that received little or no publicity.

FSD people and their efforts pioneered the whole concept of paid technical consulting services which became the backbone of IBM's revival in the mid-1990s.

I left IBM in the early 1970s to pursue a career in the corporate securities field . . . selling stocks and bonds, preparing company business plans, and writing SEC/NASD and State "Blue Sky" prospectuses for raising capital — taking young companies public.

For years the idea of writing the "FSD story" lay dormant among all the things to do "someday." The story is compelling. That "someday" has arrived.

During the 1980s and on into the early 1990s the U.S. defense budget declined significantly, procurement falling by more than two-thirds in real terms. Every defense contractor was severely impacted. A Congressional hue-and-cry for cuts in the U.S. defense budget was buttressed by the Department of Defense's own report revealing the windfall the armaments industry had been reaping since President Reagan came to power. DOD outlays to the top 100 contractors rose 28% in fiscal year 1981 to over $64 billion, more than 10% of the total U.S. fiscal 1981 budget.

One wag once remarked that since our country made war our greatest industry, business has been like sex. When it is good it's wonderful. When it's bad, it's still pretty good.

Thousands of FSD individuals made major contributions to IBM. Three stand out prominently: Bob O. Evans, Dennie Welsh and Don Estridge. Bob revolutionized the computer industry by leading IBM into the S/360 era. Later, as President of Federal Systems, he exploited the division's skilled system integration capabilities into the non-federal world such as international banking, Japanese Newspaper publishing, Japanese Broadcasting and Bay Area Regional Transportation automatic fare equipment.

Dennie Welsh's experience with FSD's NASA-Apollo work provided the spark that ignited Lou Gerstner's modern-day IBM to push into global fee-based consulting as a major business endeavor. Don Estridge's work in FSD's NASA Goddard projects established his technical credentials to take on IBM's development of the company's first Personal Computer.

FSD's greatest contributions to the defense of the United States and support of NASA's space endeavors came during the 1950s, 1960s and 1970s, the time I was privileged to say "I work for IBM."

WLR

"Every citizen should be a soldier. This was the case with the Greeks and Romans and must be that of every free state." - Thomas Jefferson

1. A WAR WAGON IN NEW YORK
FSD Roots - The Beginning

From both East and West, winds of war lashed America in 1941. In October, two U.S. Navy Destroyers, the *Kearney* and the *Ruben James* were attacked in the Atlantic by German U-boats, killing more than 100 Americans. The Nazi Wehrmacht overran Europe. The Japanese sneak attack at Pearl Harbor struck a crippling blow, pulverizing America's Pacific Fleet, killing more than 2,400 Americans and destroying nineteen ships and 188 planes.

IBM's entry into the nation's defense effort began in March of that year when the company's manufacturing facilities were offered to the government for production of 20-mm cannons. Immediately after the Pearl Harbor attack, IBM President Thomas J. Watson, Sr. turned over all the company's facilities for research and production to the government for the war effort. By the end of the war, contracts had been signed covering 38 major ordnance projects wherein IBM was either a prime contractor or a principal subcontractor.

IBM's Munitions Manufacturing Corporation in Poughkeepsie, New York, was established to produce rifles, machine guns and hand grenades for the military. IBM's dedicated workforce went to war alongside the "Rosie the Riveters" across America.

IBM punch-card units were on battlefields, keeping track of bombing results, prisoners, displaced persons, payrolls and personnel statistics. IBM machines helped break the Japanese Naval code prior to the battle of Midway and were used to help find German U-boats torpedoing allied shipping in the Atlantic Ocean.*

At World War II's end, computers became a strategic industry. During the long years of the Cold War — the U.S.'s conflict with the Soviet Union — and the Korean War, IBM entered the computer era through two major government programs: the Bombing Navigation System for the B-52 in the early 1950s and the SAGE early warning system. IBM's Vestal, New York, Engineering Laboratory was staffed with 175 engineers to test, analyze and flight test the MA-2 system. This system, called the Bomb Direction for High Speed Aircraft, was installed in a B-29 and flight tested in 1951. IBM delivered the first model to SAC in 1953. Following extensive flight tests, SAC accepted the MA-2 for installation in B-52's, awarding IBM a contract for forty-eight systems which were delivered to the Air Force in 1956.

While historians can argue over when the Cold War really began (some have said it started in Europe as early as the end of World War I) there is little to be gained from asking "who started it?" But it affected America's defense posture commencing with the Truman Doctrine announced in March 1947. General Curtis E. LeMay's SAC aircraft, 25% constantly airborne at any one time, protected America's skies during the 1950s. Buried deep in the bowels of the Midwest, near Omaha, Nebraska, shifts of controllers held at their fingertips a thousand missiles, any one of which packed the destructive force of all the bombs dropped during World War II. Americans were greatly concerned about Soviet intentions and clearly needed a network of air defense sites, thus SAGE was born.

By the end of President Eisenhower's administration in 1961, warfare was revolutionized by the atomic arsenals of the Soviet Union and the United States. Both nations had successfully tested hydrogen bombs. The U.S. had fired its first intercontinental missile more than 6,000 miles. Both

---

(* Father & Son, Thomas J. Watson, Jr.,Bantam Books, 1990)

nations were now equipped with enough missiles to engage in a massive-scale nuclear war.

SAGE (Semiautomatic Ground Environment) System was a massive military undertaking to analyze radar signals guarding the United States against Soviet aircraft. Working with MIT's Lincoln Laboratory under an Air Force contract, IBM contracted to design, manufacture, install and maintain the SAGE computers for a nationwide network of air defense sites. In May 1954, IBM established its Kingston, New York, facility, and nine hundred employees were relocated from Poughkeepsie to begin SAGE production.

By June of 1956, the first SAGE computer was shipped to McGuire AFB. In all, twenty-three systems were fabricated and installed. These became America's major surveillance and control system in the North American United States Air Defense complex.

IBM did not just enter the computer age quietly; it roared into it with the most massive computer power ever assembled. America took a tremendous jolt in confidence when, on October 4, 1957, the Soviet Union launched Sputnik, mankind's first artificial satellite; it weighed 185 pounds and traveled at 18,000 miles per hour, 150 miles above the earth. Another jolt to our security was received when Sputnik II, a much larger space vehicle was launched just weeks later by the Soviets. These two events produced paranoia and real fear that if the Soviet Union can put these types of vehicles into space it has taken the technological lead in developing intercontinental missiles, making our defenses, B-52's and SAGE, obsolete.

While SAGE could detect enemy aircraft and be used for air traffic control it was virtually useless against missiles. Sputnik destroyed America's assumption of nuclear superiority and altered the balance of power between the United States and the Soviet Union.

IBM engineers, working virtually around the clock, designed, developed, and fabricated two additional engineering prototypes — SAGE II, the AN/FSQ-32, large-scale transistorized computers which were converted to the FSQ-31 configuration and delivered to the Air Force SACCS program as government furnished equipment.

In 1955 IBM established its Military Products Division, an outgrowth of its military production effort during World War II, the Korean War and these large scale Air Force efforts. B-52 Bombing Navigation Weapons systems' production work continued, and first production prototypes were delivered to Boeing Aircraft. Technical details of these systems are contained in Chapter 11.

1958 was shaping up to be a banner year. Alaska became our 49th state, Fidel Castro began a "total war" against the Batista government in Cuba, and President Eisenhower sent army troops to the Caribbean. Vice President Nixon got "hammered" and was received with open hostility in his so-called "good will" tour of South America. It was during that August that the U.S. Army awarded IBM an R&D contract to design and develop the INFORMER System for the Army Signal Corps — a solid-state information and retrieval computer for a variety of uses at U.S. military fields and for shipboard applications.

In March 1959, the United States Air Force selected IBM to develop the 438L Intelligence System for SAC. Two months later, IBM received, from AC Spark Plug, under USAF contract, another major contract to design and develop the missile guidance computer for the TITAN II Intercontinental Ballistic Missile.

As early as 1956, IBM realized that it had to change its corporate structure to accommodate the new era and IBM's increasing size. At Williamsburg, Virginia, late that year, the company was split into six divisions plus the World Trade Corporation — IBM's overseas business organizations. Later, continued growth demanded modification, so in June 1959, the Federal Systems Division was formed. By this time, military procurement had become a significant part of IBM's business. The timing was perfect because that same month NASA chose FSD, Western Electric Company, Burns & Roe, the Bendix Corporation, and Bell Telephone Laboratories to develop its worldwide system of tracking communications for Project Mercury, America's first step to landing on the moon.

One can always say that any given year brings momentous change. But 1959 seemed to bring a diversity of change and world-wide events that could never be termed "commonplace." Three major figures were laid to rest: John Foster Dulles, George C. Marshall, and Frank Lloyd Wright, all giants on the world stage. Hawaii became our 50th state, complimented somewhat in tandem with James Michener's wonderful novel *Hawaii*. The Soviets once again startled the space race by launching a rocket with two monkeys aboard, and their Lunik spacecraft reached and photographed the moon. NASA, established just the previous year to administer scientific exploration of space, re-doubled planned efforts to launch manned space flights. (Within two years Russia's Yuri Gagarin would orbit the earth in a six-ton satellite and Alan Shepard would make the U.S's first space flight).

The new decade of the 1960s, yet to enfold, would bring a savage storm of events. First, Gary Powers would be shot down in May over the Soviet Union, initiating a crisis for Eisenhower and the nation. Next came the Bay of Pigs fiasco in 1961, followed almost immediately by the Cuban Missile Crisis, and then the 1964 Tonkin Gulf incident . . . the spark that would lead to the Vietnam War and its aftermath of national introspection. Kennedy would be shot and killed in Dallas on Friday, November 22, 1963. In 1967, Israel launched its air force against Egypt and an attack on the USS Liberty. This was followed by the attack on the USS Pueblo by the North Koreans. Death rode the global winds.

During the 1960's, there were only 5,000 stand-alone computers, no fax machines, and no cellular phones. Today, there are more than 450 million computers, (50,000 produced every day), more than 15 million fax machines and upwards of 500 million cell phones — and growing." One of the founding principles of FSD was: "The computer business is a problem-solving business, not a producer of iron."*

1111 Connecticut Avenue, Washington, D.C., had been IBM's Federal office for several decades. Originally purchased in the 1930s from a Washington bank, (complete with huge walk-in vault in the basement), it was a rather narrow three-story building sandwiched between the ageless Mayflower Hotel and the equally famous Harvey's Restaurant, often visited by U.S. Presidents. With large glass windows it showcased IBM's latest computer products, all running at top speed. It always managed to draw a sidewalk crowd.

While IBM employed some of the brightest people in the world, it also employed those with somewhat lesser talents for the necessary chores of clerking, stockroom and janitorial duties. All employees were treated equally and given every opportunity to advance upward in the company. However, some were content simply to remain doing the same job day-in-day-out until retirement. It wasn't that they were not ambitious, but that they were rather happy in their jobs. One such fellow ran the stockroom where electrical accounting machine parts and customer engineering tools were parceled out. Machine ribbons were sold for typewriters to customers and the mail room was kept quite busy before the advent of "big iron computers."

Henry Higgins (not the Henry Higgins of *My Fair Lady* fame) was the seemingly ageless dispenser of parts and supplies. Everyone knew Henry, and Henry always had a smile on his face. He stood barely five-feet three in stocking feet, had a head of unruly red hair and freckles, (a classic Irish

(* *Who's Afraid of Big Blue*, Regis McKenna, Addison Wesley Publishing Company, 1989.)

look) and would bend over backwards to be of assistance. He loved his job, he loved IBM. Henry had a big secret, though. Henry was the millionaire in the basement.

For many years, each payday at lunch time, Henry would head for a New York stock brokerage house two blocks away and buy several shares of a little known company called Haloid — which later changed its name to Xerox. Henry was fascinated about the copy wizard's prospects. For years he bought the stock, a share here, a share there. Years later, just before he retired, Henry confided to his manager what he had been doing. He said, "IBM stock was just too high priced, (in those days $500-$600 a share), and I could not afford to buy shares like I could with Xerox."

Henry retired a millionaire. His IBM retirement pay was chump change by comparison and after thirty-some years he never looked back. He drove an old Hudson, made and brought his lunch every day, never married and when they gave him a gold watch he smiled, said "thanks" and headed for those golden years someplace in sunny Florida.

"Man's mind stretched by a new idea never goes back to its original dimensions." -- Oliver Wendell Holmes

2. NEW KID ON THE BLOCK
The Federal Systems Division

On a bright sunny morning in early June, 1959, FSD opened its headquarters doors in Rockville, Maryland, a small bedroom community and farming town just north of the nation's Capitol. Why Rockville was picked for the site of IBM's future billion-dollar division no one seems to know, or remember. The town was small. Very small. Its one main street was semi-remodeled in the late 1940s. An old joke has it that the town was so small that its fire department consisted of a horse-cart and four dogs. "What do the dogs do, haul the cart?" "Gosh, no! They find the hydrant." Not only that, if someone would drive a farm tractor down main street at high noon, nobody would take notice or even stare.

This new IBM division was composed of three operating centers: the Washington Systems Center, just down the road in Bethesda, Maryland; the Command and Control Center in Kingston, New York, and the Space Guidance Center in Owego, New York. Each center operated autonomously with responsibilities to develop special-purpose information-handling systems for the Federal Government — both ground and space based. The Owego Center and the Kingston Center had extensive manufacturing capabilities, as well as their own marketing organizations to develop and sell products of advanced design.

When IBM's corporate management established a business entity, it made sure that little or no overlap in mission, product development, or marketing would impact other divisions of the company. Federal Systems therefore was given a very specific charter. The characteristics and definition of special purpose information handling systems mentioned above, had eight prerequisites:

   a. Special engineering development.

   b. Systems integrated with weapons, vehicles or mobile
tactical operations.

   c. Specially programmed and non-compatible with
standard IBM software.

   d. Hardened for shock, temperature, etc.

   e. Purchase only.

   f. Software support and service available by contract only.

   g. Price, delivery determined by contract.

   h. Non-transferable to commercial applications.

The Marketing and Service organization located in
downtown Washington, a few blocks from the White House
at 1111 Connecticut Avenue, N.W., directed marketing
programs (Joint Defense, Air Force, Army, Navy, Scientific
and Special Operations, and Civil U.S. Government agencies)
in the sales and marketing of IBM's standard product line, as
distinguished from special purpose information-handling
systems.

Within two years, Federal Systems would grow to over
9,000 engineering, systems, support and manufacturing
personnel. The formation of this new division brought
together extensive laboratories at the three centers for use in
the research, design, development and testing of its military
products.

A very savvy, long-term IBM executive, C. Benton, Jr.,
was named President. Charlie Benton worked hard, and
within months the Bethesda Center was up and running to
address governmental problems in space, undersea warfare,
intelligence and military command systems. Staffed with
dozens of PhD's in every conceivable technical discipline,
with hundreds of supporting systems engineering people,
the Bethesda Center was organized to promote FSD special
software, standard IBM computer products (where they were
needed) and special complex advanced information-handling

and control systems to meet federal agency requirements.

DEPARTMENT OF DEFENSE NEEDS

Prior to 1941 and at the start of World War II, there was no Department of Defense . . . only the Secretary of the Navy and the Secretary of War, both Cabinet members reporting directly to the President of the United States. With the advent of hostilities and world-wide responsibilities, the reaction time of the organization went from weeks to minutes. A new structure was demanded. Senior management had to be fully informed of situations developing anywhere on Earth in order to evaluate threats, convert broad action to comprehensive operational orders and to communicate command and control orders to all the commanders involved.

Today's Department of Defense structure, defined by the Goldwater-Nichols Act of 1986, has the Secretary of Defense reporting directly to the President with the chain of command to combatant commanders who command all military forces. The Chairman of the Joint Chiefs of Staff and the service chiefs, Army, Navy, Air Force and Marines, are responsible for readiness of the U.S. Military. In times of war, the DOD has authority over the Coast Guard as well.

In order to support this organization, immense quantities of information must flow directly to the top for evaluation, display and command decision. This short chain of command is facilitated by the latest in ultra-modern communications and computerized equipment. These command management systems took years to build from a semi-automated status to fully automatic processing systems.

NATIONAL MILITARY COMMAND
MANAGEMENT SYSTEMS

The Bethesda Center's depth of talent in information-handling systems led to the award of seven contracts from

1960 to 1965 to formulate development of an information system (473L) for the Air Force Chief of Staff and Air Staff, Headquarters, USAF, (the Pentagon). These contracts were to collect basic information from all Air Force commands worldwide, and to present organized facts on demand, as the information focal point supporting Air Staff decision-making. A formidable task. The main theme of the early 1950s was "short war." Simply put, a retaliatory strategic concept using nuclear payload SAC B-52 bombers. FSD's shadow force performed its work quietly, secretly and brilliantly.

The World Wide Military Command and Control System created in the days following the Cuban Missile Crisis was a system that encompassed the elements of warning, communications, data collection and processing. It was later decommissioned in 1996, but not before the National Military Command Center was established. Located in the Joint Staff area of the Pentagon, it was responsible for generating emergency action messages to launch control centers nuclear submarines, recon aircraft and battlefield commanders worldwide.

The Global Command and Control System replaced the former command system in modern times. It was an automated system designed to support situational awareness and crisis action planning. This system, again, would be replaced with a Joint Command and Control System, serving as the focal point for crisis intelligence for military operations.

Ten contracts were awarded Federal Systems beginning in 1962 for systems analysis, design and programming services related to Naval Operations management. In addition, the U.S. Navy awarded FSD a command-management contract: Mobile Operational Control Center System in the fall of 1963 for analysis and programming to support shipboard command post information storage and retrieval routines.

That same year, the Defense Communications Agency (DCA) awarded a contract to assist the command center in providing timely world-wide information in all military matters to the Joint Chiefs of Staff, the Secretary of Defense and all other persons at the highest levels of military decision making.

Federal Systems designed and fabricated one AN/FSQ-31 type special computer as a subcontractor to International Electric Corporation for the Strategic Air Command 465L Control System.

FSD's shadow force, working silently in tandem with the men and women of the armed forces, was put to the test in developing these many military command-management systems, while the systems themselves were sorely tested under extreme crisis conditions: the Cuban Missile Crisis, the Berlin Wall construction by the Soviets, and the assassination of President John F. Kennedy by Lee Harvey Oswald.

The 474L Ballistic Missile Early Warning System was the first operational ballistic missile detection radar providing an alert in case of ICBM attack over the polar region of the northern hemisphere. Radar coverage was provided at three sites: Alaska, Greenland and the United Kingdom. IBM built the first transistorized computer — four 709 TX machines for use as real-time Missile Impact Predictors. Later, in January, 1960, two IBM 7090 computers were placed in use at BEMEWS site II under prime RCA contracts. Each site had dual IBM 7094 computers for signal processing and impact prediction. Next, all three sites were upgraded with modern phased array radars and computers.

The U.S. Army's Field Army Emergency Warning System developed specific system requirements for warning all echelons of the field army of an impending nuclear, chemical and/or biological attack. This comprehensive study contract was completed in September, 1961. In conjunction with the Raytheon Corporation, Bethesda's Systems Center engineers participated in the Field Army Ballistic Missile Defense System (FABMDS), a study devoted primarily to parametric evaluation of computer techniques available in the 1965-1970 time span.

In addition to these command and control and warning systems, FSD personnel entered into numerous military intelligence and collection systems efforts, including space detection and tracking. Many of these are detailed on subsequent pages. Federal Systems support of NASA would perhaps fill a volume all by itself. Chapter 6 details much of this exciting work performed by FSDers helping to put man on the moon.

FEDERAL SYSTEMS' TALENTED WORKFORCE

Hundreds of extremely bright and hard working IBM/FSD scientists and engineering people contributed to these national defense command systems in keeping America safe . . . and they did it for almost fifty years! Creative technologies and military hardware products were conceptualized, designed, developed and manufactured by the organization's engineers. Some of the best minds in the world were hired and put to work. And work they did! Sometimes however, the best and the brightest came with idiosyncrasies and quirkiness. Much like puppies squirming in a box, they needed to be carefully handled. At Bethesda's Washington Systems Center, Joe R. Rogers was their boss.

Joe Rogers was a bear of a man with the demeanor of a very smart pussycat. His appearance was reminiscent of a middle-aged Walter Matthau, the Hollywood actor. Joe could turn on the charm as well as explode in frustration, keeping all the puppies in the box, narrowly focused on impossible bid deadlines imposed by government procurement officers. He was a computer systems engineer. He had come up through the ranks based on his smarts and his ability to lead. At times he had the look of a bemused college professor. But he was a great manager, he always got the job done, and they had him in the right place.

Personality clashes were not unknown. Time and again Joe refereed his PhD crowd, who argued over the best technical approach to satisfy military bid requirements. One PhD scientist with a chest full of degrees (physics, math, electrical engineering, etc.) had a pet monkey at home who would periodically slip his cage latch, swing from window drape to window drape, ripping them down as he flew through the living room, driving the cleaning woman crazy. The scientist would have to leave in the middle of an important conference and race home following the maid's frantic call. One brilliant mathematician, more often than not, ate his meals backwards: ice cream/dessert first, then the main course. Another astrophysicist spent long lunch hours out of the Wisconsin Avenue building at a local pub, returning in mid-afternoon intoxicated to the gills with breath that could ignite his desk. Secretaries would duck when he came down the second floor hallway. By the time he hit his office, he would be stiff enough to hang on the wall.

Government contracting in the defense industry is not an easy way to make a living. IBM's Federal Systems Division, along with every other defense contractor, always remained exposed to risks. Risks included changes in government policies and congressional appropriations, and technical uncertainties and obsolescence, including competing products. As mentioned earlier, beginning in the late 1980s and on into the early 1990s the U.S. defense budget declined significantly. These were tough years. Defense contracting always had severe limitations on profits, regardless of the type of contract. For that reason, T.J. Watson Sr. reportedly referred to FSD as IBM's "contribution to America's defenses."

As the Bethesda Center geared up in the early 1960s it hired or transferred from other areas of the corporation almost 500 systems and engineering personnel supported by 300 marketing and administrative people. This then formed the FSD's early shadow force that complimented the Owego and Kingston organizations, which grew to 10,000 before being sold in the early 1990s.

The Marketing and Service group in downtown Washington had, in addition to its primary role of marketing commercial IBM computer products to the government, the responsibility for ensuring that any equipment being sent overseas to World Trade companies had the necessary export licenses. FSD hired two highly experienced ex-military colonels, one Air Force and one Army, to interface with the U.S. State Department and DOD in acquiring the necessary clearances. Frank Wall and Fred Lippucci handled the tedious and difficult job of manhandling the volumes of administrative paperwork involved. Both World War II veterans had prior experience with the State Department as military attaché liaison officers with the U.S. Embassy and Military Advisory Assistant Groups in Italy following World War II.

Fred, over a cup of coffee, always delighted his co-workers with his tales of wartime exploits in Europe. Once a German Panzer battalion commander (with the Iron Cross hanging from his neck) personally surrendered his entire tank corps to him in the mountains north of Rome. Fred usually remarked he didn't know who was scared the most as the tanks and stacks of weapons were turned over to his small group of waiting soldiers. A number of years later, Fred joined the FSD International Group established to provide fee-based services and know-how to IBM's overseas organizations and their customers. See Chaper 8: Spreading the Word.

MARKETING AND THE
RESEARCH & DEVELOPMENT PROPOSAL PROCESS

The selling of Federal System's intellect, its highly experienced workforce and its remarkable ability to develop specialized hardware in support of the defense establishment was accomplished through the written word, *i.e.*, the R&D proposal document. Sometimes these bid packages were two and three feet tall. Whether in response to an official DOD Request for Proposals (RFP), or FSD's own in-house research's unsolicited offers, the preparation of the proposal itself was a master effort employing dozens (or more) of scientific disciplines and conceptual ideas. The decision to bid or not to bid a government RFP or to submit an idea for possible funding was the responsibility of the division's New Business Review Board (NBRB).

This board approved the project and through its members. Center and division senior management signed off on technical content, cost estimates, legal, patent and contractual provisions before its submittal. At any one time, dozens of such proposals were in process throughout the three centers. Proposal Specialists guided each of the preparations from initial ground rules meetings to final sign-off approval and delivery. Follow-on contracts, teaming arrangements or sub-contract relations with other defense contractors were similarly handled and approved by management.

This process became highly sophisticated and streamlined, resulting in FSD's obtaining a wealth of experience in its contracts with the federal government and its international endeavors.

Federal Systems experienced a reasonably high "win" rate in its R&D proposal efforts. While selective in its bidding process, Federal Systems understood its strengths, its capabilities and the market for its products and intellectual know-how. Considerable work was performed with IBM's World Trade Corporation, which came to FSD for help with its customers.

Throughout the workforce, the Federal Systems management stressed the criticality of the proposal process itself. The R&D proposal became the instrument through which most defense business was acquired and was the formal legal communication between buyer and seller. It was the essential sales tool and became the company's "silent salesman" during the government's evaluation process which formed the basis for award.

A proposal differs substantially from a technical report in that it is obligated to present the company's scheme in a persuasive manner. It must sell the idea. A proposal is persuasive writing; a report is basically informative. The obligation of reaching the reader is extremely critical in proposal writing. A report reader will overlook many shortcomings if the facts are present — somewhere — because he wants to know. No company can gamble on this occurring in the evaluation of a government R&D proposal.

FSD trained its personnel in the subtle aspects of persuasive writing, so that there was no loss of meaning, misinterpretation of ideas, intertwining of fact and fiction, nor any incomplete data which would detract from its presentation and seriously limit its chance of success.

The ability to treat the proposal without stuttering or wandering into a narrative whole from a basic premise — the company's capacity for performance — was a vital and fundamental skill. Thus, the writing took on the logic of argumentative proof derived from ancient Greek times: Ethical Proof, Logical Proof and Emotional Proof. These three are then interwoven throughout the proposal document, supporting and justifying the technical approaches taken, leading to a favorable win against the competition. Or, in the case of an unsolicited, non-competitive offer, prove the basis of the plan and why it should be accepted.

Briefly, Ethical Proof is speaking from an authoritative standpoint . . . instilling in the reader (Evaluation Board) a belief in the person (company) speaking through sincerity, well-established facts, with no contradictions and no incompleteness. Logical Proof develops the reasoning from the proposal's supporting material, *e.g.* technical dissertation, explanations, examples, statistics, analogies, and illustrations, all evidencing credibility to the reader. And, finally,

Emotional Proof, which is the appeal to the reader's self-preservation, sense of justice, good and evil, good for the national defense, entrusted employee of the government, and the responsibility in spending taxpayers' money, even the desire for power.

Federal Systems employed much of these sophisticated writing approaches in its contractual dealings with government agencies, in dealing with prime and sub-contractors and with World Trade Corporation customers. FSD won more than its fair share of contract awards for the better part of fifty years. These techniques are as applicable today as they were then. Their use for winning competitive commercial sales in today's business climate should not be ignored.

When FSD lost an award to the competition, steps were taken to immediately determine why. A critique program was instituted to determine the proposal's shortcomings. Sales and engineering personnel were dispatched to the procurement officers to find answers and report back. Answers were not always forthcoming. When they were, they were sometimes brutal. In-house "post mortems" failed to determine the exact cause until much later in many cases, when details of the winner's program and its pricing were publicly announced. Low bidder did not always win out.

An old story, dating back to almost the start of the 20th Century, had to do with Thomas J. Watson Sr. As the story goes, when he was the General Manager of the National Cash Register Company, years prior to his founding of IBM, he asked one of his salesmen, who had just lost a sale, to make a further sales call on the prospect. Tom Watson accompanied the nervous salesman and when introduced, Tom Watson merely said. "I am not here to try to sell you a cash register. I simply want to know why you *don't* want one?"

While the answer is lost to antiquity, it demonstrated a principle followed by IBM and Federal Systems. Knowing why a sale was lost is often more important than knowing why one was won! Significant future business can rest on the answer. Its one of the many reasons FSD prospered over many years as a defense contractor.

To drive home this point even more dramatically, consider the story of Thomas Edison when he was developing the incandescent light. He made experiment after experiment without discouragement or slackening of pace. After he had made fifty thousand experiments and still not found a way to produce an electric light, an assistant became discouraged and said: "Mr. Edison, we have made fifty thousand experiments and have no results." "Results!" exclaimed the great inventor with enthusiasm. "We have wonderful results. We know now fifty thousand things which won't work!" *

It is one thing to be handed a government contract because an organization is the only one with the know-how, personnel resources and the financial wherewithal to satisfy an urgent military problem such as SAGE or B-52 Weapons Systems projects. It is quite altogether different to enter into harsh competitive bidding environments to generate long-term government business. The stark reality is: how to begin? How to determine where the work will lie? The Department of Defense is a monster, with organizational tentacles stretching world-wide. How to find out, in a timely fashion, advanced weapons systems procurements and the agencies (Army Navy, Air Force, etc.) involved becomes a daunting task. While the *Commerce Business Daily* puts forth bid and contract information for potential bidders, this in no way provides the marketing information required to enter a bid for the work. When an organization first enters this world it becomes obvious that competitors have already been there, possibly influencing the specifications to be bid upon.

(* *The Speechmaker's Complete Handbook*, Edward L. Frieman, Harper & Row, New York, Evanston, London, 1952)

Advance marketing data is crucial to success. This requires "shoe work." Lots of it. A salesman with a technical background needed to be out and about learning what DOD needs are and whether his firm has the talent to respond — or for that matter even wants to respond. Without knowing what is needed he comes up short. While IBM maintained one of the largest sales forces of any business, it sorely lacked the advanced marketing data required to be even moderately successful. Recognizing this problem, IBM set about hiring personnel with such knowledge. Several retired Army and Air Force Generals were put on the payroll due to their contacts within DOD. In addition, FSD began hiring younger ex-military officers and training them on the capabilities of the three centers. Further, IBM's commercial sales force was surveyed for additional qualified people and they too were transferred in, for many had previously prowled the corridors of government while selling standard IBM computer hardware and had maintained agency contacts.

Hiring and training salesmen was but part of the marketing problem. Searching out, researching and analyzing future weapons systems procurement took a professional with considerable experience in defense marketing. Thomas G. Patterson was one such expert. In 1960, he was employed by the Nortronics Division of Northrop. FSD hired him away.

Tom Patterson had correctly forecasted changes in military buying and the "systems movement," not only within the Air Force, but in the Army and Navy as well. Tom had critiqued the Armed Services Procurement Regulations, including the new Air Force one-page proposal format in several defense marketing publications. Upon arrival at FSD, Tom set up the Federal Systems Division Marketing Staff. Several years later, after finishing the job, he moved on to RCA in Camden, N.J., with similar challenges.

One of Tom Patterson's highly acclaimed works was his annual 300-page presentation of the DOD Budget Analysis

— a complete summary statement of the Secretary of Defense on the budget. The Pentagon also cranked out a myriad of daily news releases. These were not mailed out. One had to trek to the Pentagon and pick them up. The analysis of all this data and how it might fit into the business interests of both IBM and FSD was a mighty task. Tom did it brilliantly.

Survival in the government marketplace is tenuous at best, and a company needs every tool and ingenuity that helps separate or distinguish it from the competition. The marketplace changes and evolves so rapidly that only sophisticated methods can keep astride of all the fluctuating developments in time to maneuver a company's posture to cope with them. Market research was the key. In his annual report, Tom Patterson looked at all facets of the defense budget and gave interpretations, analysis, overviews and forecasts in the areas of defense research, development, test and evaluation spending. This document became a cornerstone of FSD's successful bids long into the future on DOD needs in aircraft, missiles, electronics, ships, and under-the-surface vessels.

Tom's staff evolved into a Market Analysis Department, the formation of FSD's New Business Review Board (NBRB) and its Proposals Department, staffed with experts that shepherded the dozens of bids to the Department of Defense and other agencies of government.

One day, out of the blue, came one of the funniest guys to ever don the IBM mantle. His name was George Gerrish, an Irishman with a gift of gab that must have been blessed by the Blarney Stone. George had been the Branch Manager of IBM's Endicott Office, and when FSD was established he showed up in Washington, taking charge of the Proposals Department. It was a perfect fit for the impossible job of blending the sometimes screwball personalities of scientists and the gung-ho antics of "champing-at-the-bit salesmen" eager to make a buck. He staffed the Proposals Department with people much like himself: eager to make the written word into the gospel and generate sales for the new division. One of George's first hires was a fellow with an artist's gift for cartooning that endeared him to every salesman and manager needing to make a stand-up presentation to senior management. Brant Parker's gift was the ability to take esoteric information and ideas and translate them into easy-to-understand graphics and drawings. Brant always made it a point, however, to inform everyone that his IBM career was but a stop en route to fame in the cartooning field. Within two years Brant had sold his first cartoon, *The Wizard of Id*, to a syndicate, and he never looked back. He quit the day he signed the contract. Today, decades later, *The Wizard of Id* still brings merriment to the comic pages of hundreds of newspapers. George was sorry to see him go. Brant's shoes were hard to fill.

George Gerrish never made an enemy. He was well liked throughout his career in FSD moving from Proposals Manager to Systems Sales Manager for the division's "black ops" work with DOD, CIA, NSA and the FBI. George left IBM much too soon. On a sunny day in the early 1980s thousands of people jammed the cathedral-like church in Chevy Chase, Maryland, for his funeral. Police stopped traffic as the crowd spilled out into the streets. George had died walking his dog one night. FSD lost a highly successful manager and the world had lost a "Danny Kaye" personality.

(I'll bet that if you listen closely, you will hear the laughter up on high.)

THE KINGSTON, N.Y., GROUND CONTROL CENTER

In the 1960s the FSD Kingston organization, while developing big processors and display systems, developed special systems for space and military applications. One such project was the IBM 2250, used for the Gemini flight simulator. This center developed and delivered an air traffic control system for the FAA at Atlantic City, N.J. In 1969, the center developed a new control method for data processing used on the Apollo 11 mission.

A Bright-Tube Display Console was fabricated and delivered to the MITRE Corporation in early 1962 and four additional ones fabricated for Project WS-239A, a highly classified Air Force satellite project.

The Kingston Center was awarded a contract by the Army Corps of Engineers to develop hardware and a computer program to demonstrate the production of topographic maps and contour manuscripts by digital computer techniques. Following completion and demonstration, the equipment was delivered to the Research and Development Laboratory at Fort Belvoir, Virginia.

The Center designed test equipment for the B-70 Guidance and Navigation System and also supplied factory and field test equipment for the SAGE computer, the SAC Control Center computer, and the Fieldata INFORMER. The Fieldata INFORMER was designed, developed tested and delivered for use in Army field environments. This system, composed of a general purpose central processor capable of 30,000 to 40,000 instructions per second, was housed in a standard S-109 Army shelter for use on a two-and-one-half ton truck. This ruggedized mobile system was developed for many applications, including information retrieval, logistics support, personnel, fire control and automatic map compilation. The technology used was pulse magnetics. The INFORMER system underwent acceptance tests at USASRDL (U.S. Army Signal Research & Development Laboratory) Fort Monmouth, N.J.

The Canadian AN/FSQ-7 SAGE system was a deep underground system in which Kingston designed and developed modifications and special environmental testing to ensure no problems would be encountered in an underground placement.

The Command and Control Systems group within the Kingston plant was staffed with engineers with experience in designing, developing and fabricating large and medium-sized high-speed mil-spec digital computer systems as evidenced by the SAGE program.

Battlefield technicians must cope with bad weather, worse terrain and loss of men and equipment during any hot conflict. Field commanders must employ all their resources most effectively and look to advanced methods for directing attacks and the most advantageous use of defensive fire power. Commencing with the FABMDS anti-missile system study performed by FSD personnel in 1960 and 1961, various groups within the Washington Systems Center analyzed the utility of computers to support military tactics and the development of specialized hardware for this mission.

The Field Army Ballistic Missile Defense System (FABMDS) was a joint study performed with the Raytheon Corporation to explore the feasibility of a field army employing tactical data processing by machine methods. Federal Systems' portion of the contract was devoted primarily to parametric evaluation of computer techniques available and applicable in the 1965-1970 time span. This system analysis embraced endo-atmospheric trajectory predictions, decoy characteristics, computer size and high speed real-time methods of calculating.

The ARPAT project, a ballistic missile defense system, was a joint study with Raytheon in 1961 and 1962 for the Advanced Research Projects Agency (ARPA). This effort outlined a fire control data handling system to provide defense against ballistic missiles during terminal flight.

An Army air defense system of the 1970s was designed through a joint effort, again with Raytheon, to demonstrate the feasibility of defending field armies against both aircraft and ballistic missiles. The short reaction time necessitated a complete automation of controls and decision mechanisms. FSD analyzed weapon-control subsystem requirements and designed high-speed reliable data processing networks in the field environment including suitable self-propelled shelters or those transported by aircraft. This AADS-70s project combined particular skills and talents of both Federal Systems and Raytheon.

Working with Bell Telephone Laboratories, FSD's shadow force supplied programmers and program analysts on location at Whippany, N.J., for the Nike-Zeus Phase II Computer and evolution into the Nike-X. Simultaneous with this work FSD personnel, under subcontract to Raytheon were engaged in a U.S. Navy contract "Advanced Surface Missile System" (ASMS) with several other contractors: Vitro Corporation, Bath Iron Works, ITT-Gilfillan, and Gibbs & Cox. The purpose of this program was to (1) study and recommend alternatives for the integration of weapons direction and fire control centers for the existing NTDS Naval Tactical Data System to improve a ship's capability to defend itself or the fleet against attacking aircraft or missiles; and (2) design an integrated defense system for future naval vessels.

FSD designed, installed, checked out and maintained a target acquisition system for the U.S. Naval Ordnance Test Station at China Lake, California. This application of aerial surveillance processing employed an IBM 7090 computer to direct the pointing of optical tracking devices on the missile range as five independent radar stations send target position information in real time to a central computing station.

Under contract to the Naval Command Systems Support Activity (NAVCOSSACT), commencing in 1962, FSD engineers provided analysis, programming, installation and maintenance services under four contracts to Navy Operations and Plans. These personnel were in residence at CINCLANT and CINCPAC locations and in Washington, D.C.

A number of purchase order contracts funded the Air Force 461-L program to Lockheed and IBM. Lockheed enlisted FSD support in developing the ground-handling mechanics of a missile defense satellite system. This Air Force system incorporates aerial surveillance data processing in conjunction with orbiting satellites that detect intercontinental missiles at the time of first launch. The FSD simulation and programming staff was located on site at a Lockheed Laboratory.

The Bethesda Systems Center also was awarded an AF contract to study and analyze concepts relating to satellite-carried aerial surveillance processors. Under an additional contract with the Bendix Corporation, FSD produced computer programs for processing data from a massive, electronically steered array radar at Towson, Maryland.

The SPADATS, Space Detection and Tracking System, encompassed the development and installation of a phased-array radar system for maintaining accurate and up-to-date information on all artificial earth satellites. Raw radar data was processed at point of origin (Eglin Air Force Base) and then forwarded to NORAD headquarters at Colorado Springs. FSD provided the computational interface between the Bendix FPS-85 radar and the government link. Installation of IBM 7044/1401 systems at the radar site and an IBM 4554 display unit under a separate contract at Eglin AFB. This was a $2.1 Billion contract under Bendix, which began in the third quarter of 1962.

IBM 1311 MIL Disk Storage Drive — FSD developed the 1311 Mil File in the mid-1960s as the solution to the mass memory requirements of military tactical data systems. This memory file is a militarized version of IBM's standard 1311 Disk Storage Drive, re-engineered to function reliably in extreme tactical environments.

By the 1980s, Federal Systems scope of activity and its experience built upon experience, project by project, over several decades, expanded dramatically. Many ground-based problems emerged that could be addressed with standard IBM equipment, but which now required FSD's singular expertise with real-time systems and computing technologies and its ability to manage very large-scale programming efforts. Examples were ground control of positioning satellites, air traffic control systems for the FAA, and complex problems of civil government agencies such as the U.S. , Post Office and Justice Department.

During the late 1980s, IBM re-designated the Federal Systems Division as the Federal Systems Company, in final recognition of its organizational flexibility to address a much larger customer base, including military, civil, domestic and international non-governmental commercial enterprises.

Until then, Federal Systems was an organization virtually indivisible from its parent IBM . . . an incredibly talented, semi-clandestine, shadow force serving America's interests.

Managing and operating an organization of this size required a support staff of hundreds: clerks, secretaries, managers, junior engineers, accountants and purchasing officers, printing and publishing people and the legions of machinists and technicians to design and build equipment.

In 1970, one of many long-time IBM/FSD employees retired. His name was Bernard L. Boteler Sr. He had spent almost his entire working life with IBM. In fact, he pre-dated IBM, having worked in Washington, D.C., for the Computing Tabulating Machine Company — one of the companies Thomas J. Watson Sr. joined together to form International Business Machines Corporation in the 1920s. If Bernie wasn't the first employee, he sure was close to it.

Bernie started work as a teenage card-printing operator at the New York Avenue Card Plant, fabricating cards for the U.S. Census Bureau. Later, he served as the general office manager. In the 1930s the U.S. Government began paying all employees with blue, punch-card payroll checks. These were printed in a special area of the plant, since the card stock itself, once printed, became negotiable and strict controls were mandated . . . it was like printing money without a signature. Bernie was swept up into the new IBM company, eventually serving as IBM liaison to the Veterans Administration and the Census Bureau during World War II.

Bernie moved on to FSD as a Senior Purchasing Officer, retiring in 1970 after fifty years. Bernie died in January, 2003, at the age of ninety-seven. He probably had the longest continuous run of any employee ever hired. His efforts over many decades helped build the IBM Corporation and Federal Systems. IBM had many long-term employees, but few that pre-dated Tom Watson's dream nearly a century before.

There is no doubt that Federal Systems enjoyed a fine reputation in the defense industry. It was not however, one of the biggest, not even in the top ten when measured in dollars spent on national defense. Companies such as McDonnell Douglas, United Technologies, General Dynamics, Boeing, Lockheed, Hughes, Raytheon, General Electric and others annually received the majority of defense outlays. After all, they produced military aircraft, nuclear submarines, missiles/space systems, ships and army tanks — winning multiple billions of dollars in annual government defense funding.

In fact, in the early 1980s, Department of Defense outlays to the top 100 prime contractors totaled $64.7 billion. When these figures were released, even Republicans were crying for cuts in the U.S. defense budget. As the 1980s drew to a close, fiscal year budgets declined significantly, producing major changes in corporate business activity. While FSD's military business declined as well, major contracts with other agencies of federal and state governments picked up some of the slack. By 1993, when FSD was purchased by Loral, Federal Systems turned over a sales backlog of more than $3 billion.

What set Federal Systems apart was its unique product mix of people, computer/programming prowess and specialized information-handling equipment which not only could be bid direct to government agencies, but proposed to other defense bidders as well, based upon the specific nature of the weapons systems procurement. In fact, teaming, joint bidding and acting in a subcontractor role to other "competitors" rather than as a prime systems contractor produced significant work and revenues to Federal Systems over the decades.

FEDERAL SYSTEMS LABORATORIES & FACILITIES

Federal Systems established extensive laboratories at its three centers equipped with modern equipment for use in research, design, development and testing of products for military use.

These facilities included a simulation laboratory, a steller-inertial laboratory, environmental test laboratories, analysis and computational centers, thin-film and cryogenics laboratories, flight evaluation and miniaturization laboratory equipped with Swiss machine tools, various chemical and physical research laboratories and a communications laboratory.

The Kingston, New York, Ground Control Center had an optics laboratory; a crystallography laboratory; a chemical, metallurgical and spectrographical laboratory, a controlled atmosphere laboratory containing material analysis equipment including Jarrel-Ash mass spectrometer and X-ray and electron diffraction equipment. In addition, a fully equipped photographic laboratory was installed.

Owego Space Guidance Center was equipped with numerous facilities: a simulation laboratory; a computation center with the latest models of IBM's large computers; a microminiature laboratory; a super-clean area, a failure analysis, steller-inertial, Applied Physics and Flight Test Laboratories. The last, which performed functions necessary for the flight testing of products, was located at the Broome County Airport near Owego, New York.

Liberty is always unfinished business.
- Anonymous

3. IF HANNIBAL HAD ONLY HAD A JEEP
Intelligence & Collection Systems

If Hannibal had only had a Jeep was a brilliant thesis statement, part of a "white paper" written by FSD's Dr. E. Castruccio in response to the U.S. Air Force Research and Development Command (RADC) on future war in space. He explored the hypothetical ramifications of future battles in a "star wars concept." He rightly concluded that the most important aspect was intelligence. Information about the antagonist was tantamount to survival. He used Hannibal's crossing the Alps as the basis of his response.

Hannibal was a Punic leader, only twenty-eight years old and in his prime, winning battles with his brains, his bravery, the skill of his espionage and the subtlety of his strategy and tactical surprises. Had he had the ability to quickly search out and identify the problems in crossing the Alps with his elephants, history would surely have been altered. A highly mobile platform collecting intelligence of the way ahead would have saved half his army in 218 B.C.

Extraordinary hardship was encountered in getting his elephants through narrow and precipitous mountain passages. His army climbed for days. Reaching the summit, he found it covered in deep snow. With just several days of rest, Hannibal began his downward march through passes steeper than the ascent and over paths buried in landslides and paved with ice. So many horses, elephants and men were lost that when he reached the plain his army had been reduced to less than half of the force he had left with from

New Carthage four months earlier.*   Dr. Castruccio's treatise went further in identifying how such intelligence would be used, the speed at which it must be acquired and the futuristic weapons response.

Intelligence reduces risk by identifying adversary capabilities, vulnerabilities and intentions in the battle-space. For instance, during the Cold War, the Navy established Ocean Surveillance Information Systems which focused U.S. Naval intelligence efforts on the Soviet Navy. It goes without saying that every field commander who knows the enemy's location and the types of forces being deployed enjoys a great tactical advantage. Awareness is key to battlefield dominance.

After World War II, until the creation of DIA, the Defense Intelligence Agency, the three military departments collected, produced and distributed their own intelligence for individual use. This proved to be duplicative and costly, as well as ineffective. The DIA was established in 1961 to solve this situation. During that summer, Cold War tensions flared over the Berlin Wall and the 1960s brought on China's detonation of an atomic bomb, fighting in Cyprus, the Tet offensive in Vietnam, the missile gap between the U.S. and the Soviets, the Six Day War between Egypt and Israel, North Korea's seizure of the USS Pueblo and the Soviet invasion of Czechoslovakia. DIA earned its stripes during its first decade in business.

Since its formation, Federal Systems has been in the forefront assisting the U.S. military in intelligence and collection systems. As early as March 1959, MPD/FSD began a systems analysis relative to intelligence data handling applications for the 438L SAC subsystem. Under a series of follow-on contracts, covering more than five years, FSD proved instrumental in supporting the growth of warning systems under the USAF/DIA Data Handling System (IDHS). As a subcontractor to RCA, the Bethesda Center participated

---

(* *The Story of Civilization*, Will Durant, Simon & Schuster, 1950)

in Project 466L, Electromagnetic Intelligence System, wherein FSD provided programming efforts utilizing extensive IBM computer equipment. From 1962 through 1963, a joint IBM Navy team designed and programmed a generalized information system for the Fleet Intelligence Center Europe (FICEUR) in Morocco. A follow-on contract adapted the initial system for use by the Fleet intelligence Center Pacific (FICPAC) as well as for the Atlantic Intelligence Center.

The U.S. Army Scientific & Technical Information Functions and Activities Data Bank contract, from the Army Research Office, established a mechanized data bank to process, maintain and update data gathered in on-site surveys of technical information.

Under an Air Force Contract, FSD systems personnel teamed with the University of Pittsburg's Health Law Center to devise a total text information retrieval system for the USAF Accounting & Finance Center. Project LITE promoted the advantages of high-speed, random access, input-output equipment for application to large data base systems.

The SALAD Systems Analysis — Low Altitude Detectability — contract with the U.S. Air Force's Wright Patterson Air Force Base was a study performed by Federal Systems analysts to determine the survivability of a future air vehicle of advanced design and performance in penetrating hostile territory at high speeds. (Could this have been the earliest planning for stealth aircraft)?

FSD's Washington Systems Center personnel conducted numerous studies under military contracts as well as under its own internal IRAD (Independent Research & Development) program in the intelligence and collection systems area. Many projects were (and remain) classified. The preceding pages recount just a sampling, which exemplify the breadth of experience Federal Systems personnel acquired during the several decades of service to the government.

In addition to supporting military and civil agencies

using true intelligence data, the Bethesda Center assisted agencies in developing systems for storing and retrieving non-intelligence information. For example:

** Engineering documentation
** Legal decisions and codes
** Identification and medical records
** Geodetic surveys
** Bibliographies and photographic indexes

Federal Systems personnel well understood that the market for collection and intelligence systems included design, development and implementation of information systems (both digital and pictorial) whose operations necessitated file-oriented data bases. This market, which FSD successfully penetrated, encompassed three broad areas:

a. Intelligence Processing — those data processing and system development activities oriented to particular problems of the intelligence community.

b. Intelligence Collection — systems directly concerned with the nature of sensors and with immediate interpretation or conversion of sensor data for later exploitation in intelligence processing systems.

c. Documentary Data Handling — systems activity for non-intelligence federal agencies whose functions necessitate information retrieval systems that employ techniques similar to those used in intelligence data processing.

One of the Bethesda Center's strengths was its ability to design and implement systems that operated in some combination with the above areas. It is axiomatic that as intelligence moves closer to the means of data collection, the sensor becomes all-important with respect to redundant readout. Some sensors, by their very nature, collect more data than is needed or even usable by the central processor. (Radar, for example can collect large amounts of extraneous or redundant data.) Redundancy reduction at the sensor can operate to conserve communications capacity and to improve the quality of data being sent to the processor.

IDHS/ACIC PHOTOGRAPHIC DATA PROCESSING

Under an Air Force contract, Bethesda Center personnel analyzed, designed, developed and tested a computer-oriented system for accessing the holdings of the USAF Aerial Film Library. The primary purpose was to provide a data processing capability affording a more timely and adequate service to film library users. This contract was one of many that paved the way toward automated documentary film handling.

The Aeronautical Chart and Information Center (ACIC) maintained a library of aerial photography which included cut film, roll film, photographic prints and documentary photography. Through FSD efforts, retrieval of this data was effected by an integration of personnel and equipment skills and techniques. The final system allowed the accession of individual frames of selected photographs, those in the library, and aerial photographs acquired at a later date.

Indexing and retrieval procedures operated in conjunction with an IBM 1410 data processor. The operating capability provided a computer-oriented index to eight million frames of aerial photography, with a gross index of thirty million frames.

It was during Eisenhower's administration that the military put new technologies to work in defending the United States, especially the classified spy satellites. As these new intelligence gathering tools came into operation the problem became greatly amplified in the more advanced satellites of the 1980s and 1990s. The problem confronting the intelligence community was the making use of all this information, rather than its collection. The amount of data was overwhelming, and the problem for analysts was simply to identify the urgent and important data and then to make sense of it.

Today, these satellites remain highly classified. It can be

said, with good solid intelligence about an opponent, the United States can avoid wasting its defensive resources. Thus, through to the end of the twentieth century, America's intercontinental missiles, early-warning radar systems such as SAGE, the U-2 spy planes and spy satellites became our first line of defense.

Much information comes from communications,
and much communication comes from misinformation.
- Anonymous

4. FROM THE DEPTHS TO THE ETHER
Marine Systems & Communications Systems

Virtually since its beginning Federal Systems employed personnel in disciplines especially pertinent to marine information handling. The earth's water bodies are foreign volumes that posed special problems relative to investigating their economic and politico-military worth. In 1964, FSD had over thirty marine systems technology experts in:

1. Hydrodynamics of oceans and atmospheres
2. Applied mathematics (also information theory)
3. Signal processing, measurement, and recovery, and
4. System engineering, synthesis and simulation.

The capabilities of this marine group were applicable to research and analysis in underwater sound propagation and signal processing, acoustical systems engineering for shipboard or stationary marine installations, control and oceanographic systems involving instrumentation development, signal conditioning, analysis, acoustic array design, propagation analysis and underwater communication.

FSD conducted studies in the mid-sixties of integrated submarine control (SUBIC) problems under contract to Electric Boat Company. GALS, Gravity Anomaly Location System, was developed under contract to the Woods Hole Institute to provide means of measuring absolute gravity at sea. As the result of this program's successful operation, the contract was expanded and the equipment underwent extensive testing on the research vessel R.V Chain. Subsequently it was declared operational, performing extensive surveys as part of the International Indian Ocean Expedition participated in by the United States.

Artemis-Multiplexer & Bulk Delay I
(Acoustic Beam Forming)

In order to process data being received by the Artemis hydrophone array, it was necessary to multiplex and transmit data via microwave from a Texas Tower at Argus Island to the laboratory in Bermuda. IBM constructed, installed and tested the multiplexer, whose information when received in Bermuda, was then formed into beams for analysis. A bulk delay was designed and built to introduce proper delays into the hydrophone responses to take into account the location of hydrophones and the shape of the acoustic wavefronts arriving at the array. This system successfully operated in Bermuda and as a result, Hudson Labs awarded the company a follow-up contract to build an even larger system.

Autec-Acoustic Tracking System

A problem area in the acoustic Noise Measurement Range was to find a satisfactory means for ensuring that test ships at the David Taylor Model Basin passed at a predetermined distance from the noise measuring hydrophones along a desired track. Federal Systems was awarded a contract to design, build, test and place into operation a system to accurately track and guide test vessels, including submerged submarines. Final tests were successfully completed in 1965.

AROD Ocean Platform Feasibility

FSD Marine Systems personnel studied the problems of mooring radio transponders on the ocean's surface at locations where the depth may exceed 20,000 feet. In reporting the final results, personnel cited limitations of present buoy designs.

In-House Research on Sonar Propagation and Signal Propagation, ongoing for more than four years since the formation of the Marine Systems Department, resulted in many significant achievements: the development of a digital computer based system for the analysis of acoustic signals and the simulation of sonar signal processing systems using actual acoustic signals; the development of a general purpose computer program for the design and analysis of acoustic arrays, and the design of a conformal array beam forming system permitting beam balancing for passive detection for classification purposes of a bow array for submarine sonar. This effort employed a dramatic new signal processing technique — the development of a matched filter classification system and the development of coherent spectrum analysis techniques for use in passive sonar classification. A further contract with the Bureau of Ships demonstrated the effectiveness of this technique when later incorporated into the AN/SQS-23 sonar system.

A great deal of this original work at FSD resulted in the development of FSD's Advanced Signal Processor for sonar applications, designated by the Department of Defense as the DOD Standard signal processor, AN/UYS.1. Sonar applications include operation in long-range patrol aircraft, fleet protection helicopters and destroyers and in submarines.

FSD's Marine Department personnel addressed meteorology, especially military meteorology, satellite meteorology, and atmospheric research and analysis. Contracts were awarded by the U.S. Weather Bureau in many technical areas. Personnel participated in and supported several manned and unmanned civil and military satellite studies conducted by FSD's Owego's Space Guidance Center. (See Space, Chapter Five)

Communications Systems

Federal Systems established its Communications Systems

Center in the fall of 1960 to meet the federal government's requirements for special communications systems and equipment. While much of this work was classified, it encompassed the development of high-speed, high-data-rate, digital data modems, error detection and correction devices and techniques, switching systems, adaptive HF radio systems, signal processing circuitry, multiple access communications systems, and applied research in laser technology. Specialized hardware was designed and developed, including a FLEXCODER for on-line testing and demonstration of error detection/correction code applications; switching systems in networks to meet special complex government specifications, procedures, programming, formatting and cryptographic message handling.

The center was staffed with more than 100 skilled scientists, engineers and specialists in multiple communications disciplines. One of the earliest contracts was with DCA, the Defense Communications Agency. This contract called for equipping four DCA centers with advanced information systems providing minute-by-minute status information with locations in Colorado, Hawaii, France and the National Defense Communications Center near Washington, D.C. This system was a vast complex of lines, trunks, and switching centers comprising over 6,300 circuits spread through 73 countries around the world. IBM standard computer equipment was installed at each center.

Contained within the FSD communication organization was a Space Program office concentrating on telemetry-communications oriented satellite payloads; data compaction, and signal analysis. A contract with the U.S. Army's Fort Monmouth, N.J., command was awarded to FSD in the area of television compaction for reducing channel requirements for military transmission.

The Communications Systems Center designed and installed complex message switching systems combining standard IBM equipment with unique special devices in networks to meet stringent government specifications for procedures, programming, formatting and cryptographic message handling. An example of this combined systems approach was the design and implementation of an automatic status and traffic reporting system for the Area Control Centers in the Defense Communications Global Defense Communications System.

Special purpose signal processing is essential to many aerospace and tactical systems. Multiple Access Communications Systems that operate without conventional switching equipment promise great advantages in terms of mobility, survivability, and logistical simplicity. The center developed a precision phase-lock loop receiver designed for NASA's Saturn program, and a range and range-rate navigational system was developed using this phase-lock-loop technique.

RADC awarded a contract to investigate switching systems, calling for investigating the feasibility of various techniques in the design of four-wire circuit switching equipment that would provide full-duplex, real-time switching. The microminiaturization approach to component techniques, switching methods (space division, time division, modulation techniques and bandwidth studies) formed the core of this work. This FSD organization was staffed with personnel with over 159 years of military service and more than 100 years of industrial experience directly related to telecommunications systems and data systems.

Robert P. Crago, a senior systems engineering manager who had helped build SAGE Systems in Kingston, New York, was promoted to manage this new Communications Center. Bob personally hired and staffed the department, and immediately set about assigning his engineers to a number of IRAD in-house study efforts in the fields of circuit switching systems, ionospheric transmission studies, the use of coding for error control, voice data multiplexing and principals of adaptive network design. These tasks were augmented, as personnel came aboard, by studies in H.F. radio compatible modems, space antennas, advanced modulation techniques, secure voice transmission and bionics. Needless to say, this new communications organization, with its talented engineers and scientists, seldom lacked for work. In some cases it "loaned" its personnel to other FSD departments under government contract coverage, when specialized communications knowledge was needed.

The mark of talent is to do the possible with ease; the mark of genius is to do the impossible with difficulty.
- Anonymous

5. DEFENDING AMERICA'S SKIES
Military Aerospace Systems

While America's National Aeronautics & Space Administration (NASA) has focused its resources on man's peaceful exploration of space (See Chapter 6), the United States Air Force as well as the U.S. Army has concentrated its forces on defending the U.S. from all enemies.   While NASA's endeavors receive extensive publicity, the military maintains a low profile and secretive nature regarding its satellites, missile detection, space detection and tracking, and ground-based missile defense.

It is not generally known, but even NASA's flights at times carried out classified projects for the military and the USAF had control of the launch facility at Cape Canaveral. Each of the centers of the Federal Systems Division were firmly identified with aerospace flight activities since inception. A difference in mission separated the Washington Systems Center, the Command Control Center and the Space Guidance Center. While Owego's Space Guidance Center emphasized the special hardware for on-board flight control, the other two centers concentrated on implementing ground-based flight control systems.

B-52 Bombing Navigation System

Chapter One briefly described IBM's early work on the Bombing Navigation System AN/ASQ-38 for B-52's during the early 1950s. Design and development work was performed by IBM's Airborne Computer Laboratory (ACL), forerunner to its Military Products Division. Prototypes were tested by IBM engineers at Broome County Airport,

Binghamton, N.Y., and operational flight tests performed by the B-47 jet aircraft. The AN/ASQ-38 Weapon Control System was an offensive system. The B-52 was developed by the Boeing Aircraft Company under a prime Air Force Systems Command contract. Later, Federal Systems had overall management coordination responsibility for the development and integration of weapon system AN/ASB-9A. Put simply, this complex system calculates and compensates for such variables as air speed, altitude, drift, Earth's rotation, bomb ballistics, roll, pitch and yaw to guide the B-52 to a predetermined release point over the target drops the bomb within a 1,000 feet radius of the target, photographs the damage and guides the aircraft back to the base. (See Chapter Eleven for technical specifications).

B-70 Bombing Navigation Missile Guidance Sub-system

This System is a modification of the ASB-9 system and contains analog computers for solving the navigation and bombing problems and presentation equipment for displaying radar video intelligence obtained from search radar equipment. These modifications were carried out by FSD's Owego facility under Air Force prime contracts.

The Space Guidance Center also developed the AN/ASQ-28(V) System, a completely integrated subsystem designed for Mach-3 environments for use in a supersonic, manned weapon system. SGC also developed a back-up computer, the AN/ASQ-28(v) Emergency Digital Computer (EDC) for the B-70 System. (Technical details of these systems are contained in Chapter 11.)

Ballistic Missile Defense

Federal Systems established a Space Systems Department to study ballistic missile defenses systems . . . more properly called anti-missile missile systems. An early contract ARPA's

(Advanced Research Projects Agency) of DOD was a sub-contract under Raytheon for feasibility studies of the fire-control subsystem Terminal System. This work involved various studies of trajectory predictions, error analysis, the dart assignment problem, data processing techniques, fire-control subsystem configurations, and terminal guidance and control techniques.

In addition, FSD provided programmers and analysts to the Bell Telephone Laboratory in Whippany, New Jersey, to aid in doing operational and development simulation programs for the Nike-Zeus anti-missile missile system.

Defensive Satellite Ground Data Processing

A number of subcontract purchase order contracts funded the IBM 461L Air Force Program during the mid-sixties. Lockheed Missiles & Space Corporation enlisted FSD support in developing the ground-handling mechanics of a missile defense system. The Air Force 461L system incorporated aerial surveillance data processing in conjunction with orbiting satellites that detect intercontinental missiles at the time they are first launched. The ground system must perform the entire task of position determination, target tracking and impact predicting, in addition to maintaining cognizance of all new targets.

Computer Control of Target Acquisition

Federal Systems designed, installed, checked out, and maintained a target acquisition system for the U.S. Naval Ordnance Test Station, China Lake, California. Five independent radar stations send target information in real time to a central computing station (an IBM 709).

Space Detection & Tracking System

Under a RADC (Rome Air Development Command) contract, AF Project AN/FPS-85, Federal Systems personnel worked under a Bendix Corporation purchase contract. This system encompassed the development and installation of a phased-array radar system to maintain accurate and up-to-date information on all artificial Earth satellites. Raw radar is processed at the point of origin (Eglin Air Force Base) and forwarded to NORAD headquarters at Colorado Springs. As subcontractor, IBM provided the computational interface between Bendix hardware and the government data link reaching out to the NORAD Space-Track Control Center. In addition, under this $2 million dollar contract, FSD's shadow force developed computer programs that control the radar's functions of surveying and tracking. This large programming effort comprised 75,000 instructions, written primarily in FORTRAN language.

Air Force Satellite Test Center

Federal Systems, under contract to Lockheed Martin, performed system analysis and program design for the Air Force Satellite Test Center, Sunnyvale, California. This center was responsible for the testing of all satellites in the WS-239A, SAMOS, DISCOVERER and ADVENT series and other classified military R&D satellites. The WS-239A missile defense system was an early warning system based upon the immediate detection of missile launches by infrared sensors housed in orbiting satellites.

OAO The Orbiting Astronomical Observatory

The OAO was a scientific satellite which traveled at 18,000 mph in a 500-mile orbit, collecting observations of the heavens outside the distortion of the earth's atmosphere. As the satellite circled the earth, IBM equipment starts and stops each experiment, tells OAO's sensitive telescopes where and when to look, and checks for malfunctions. FSD's Space Guidance Center, Owego, under contract with the Grumman Aircraft Engineering Corporation designed, built and evaluated five models of the Primary Processor and Data Storage Equipment (PPDS). The first OAO was launched from the Atlantic Missile Range (Cape Canaveral) in 1963. This on-board processor utilized quad-redundant components. This unmanned vehicle operated error free for four years in orbit. (See Chapter 11, for technical specifications).

DACAPS (Data Collection & Processing System)

Under Air Force Contract AF-08(635)-727, FSD personnel conducted studies that were used to design an integrated data handling system to collect and process data arising from aerospace research experiments conducted on the Eglin Gulf Test Range.

TITAN II ICBM

The Titan II was a second-generation missile reflecting certain advantages over earlier missile systems, including all-inertial guidance, increased range and the use of storable fuel which permitted the missile to be maintained in a "cocked" condition within a silo. FSD Owego provided the airborne digital computer for the Titan II guidance system as well as the check-out and launch equipment for monitoring the missile's continual state of readiness. The Titan II (ASC-15 computer) was FSD's first on-board computer in the missile and space field. This computer was later used in both Saturn I and IB programs.

In less than one year a prototype was delivered and additional computers also delivered for many successful Titan firings. The computer weighed less than ninety pounds and its miniature components were tough enough to resist all deterioration. The computer can withstand the severe vibration and jarring shock of Titan underground missile firing and acceleration. FSD's Owego's engineers instituted manufacturing process concepts, up until that time, previously unknown in the industry.

TIROS I (Television Infrared Observation Satellite)

The Tiros I was a 270-pound satellite. Its mission was to televise cloud cover, storms and weather from an orbit 468 miles above the earth. FSD engineers were responsible for an entirely new electronic computer program that determined when the television camera lenses were aimed at the sunlit portions of the earth. When this was determined, photographs were taken, recorded and relayed to U.S. receiving stations.

The task of deriving meteorological information from the tens of thousands of photographs received from TIROS, the weather observation satellite, led to a contract from the Air Force Geophysics Research Directorate for an image processing feasibility study. The purpose of the program was to use image processing to correct the photographs for tilt and to convert the photographs of the earth into a two-dimensional Mercator map projection.

TIROS 35-mm photographs were scanned with the modified sterocomparator used in Phase II of DAMC (Digital Automatic Map Compilation) contract with the Army Corps of Engineers. Before image processing could be accomplished, it was necessary to express mathematically the relationship between the old and new data formats, and to establish the relative orientation of the TIROS camera to the earth.

ATHENA REENTRY RESEARCH

Under a U.S. Air Force contract (the USAF Ballistic Systems Division), FSD, in cooperation with the Aerospace Corporation, participated in missile reentry simulation at White Sands Missile Range in New Mexico. FSD had established a Western extension to support government agencies operating in or near Pacific states. The work on

Athena involved the better understanding of reentry phenomena. This knowledge had an immediate application to increased penetration of ballistic missile nose cones.

This research involved the launching of an Athena booster at Green River, Utah and causing impact at the White Sands Missile Range about 420 miles away. IBM/FSD engineers were in-residence at WSMR. Three computer systems were utilized: an IBM 7094 and two 7044's. Please refer to Chapter 9 for an in-depth review of Owego's aerospace computers and their technology.

LAMPS (Light Airborne Multipurpose System)

ASW Helicopter Systems integration is at the heart of the military's evolving requirements in aircraft, ship, and ground vehicle defenses. The challenging task is to effectively integrate weapons, platforms and sensors to create a seamless network of defense capabilities. In 1974, FSD Owego was awarded this Navy contract as the SPC, Systems Prime Contractor for the LAMPS weapon system providing the fleet with ASW, anti-surface and airborne early warning protection. The system performs four functions: (1) Detection, (2) localization, (3) classification, and (4) attack, embodying the integration of the parent ship, frigate, destroyer, cruiser, and the manned aircraft, SH60B Seahawk helicopter.

The AN/BQQ5

The Navy's first digital submarine sonar system for onboard signal processing and combat was designed and developed by FSD Owego scientists and engineers. This highly classified program was a significant breakthrough in warfare capability. The U.S. Navy also awarded FSD a $683.4 million contract for six sets of advanced submarine combat systems in December, 1983.

Space is being conquered by man while man is being conquered by time.
- Anonymous

6. BIG BLUE INTO THE BLUE
Space — NASA

A great deal has been published over the past fifty years about NASA's manned and unmanned missions into space. The entire world knows about America's successful moon landings and its tragic failures. Since NASA's inception in 1958, the IBM Corporation's extensive resources were deeply committed to supporting NASA's space flight programs: projects Mercury, Gemini, Apollo, Space Shuttle, Skylab and the International Space Station. IBM Archives present a chronological history of Space Flight, actually commencing in 1944, with its design of the Automatic Sequence Controlled Calculator at Harvard University, to be used by Navy scientists to prepare ballistic tables.

While much publicity accompanied IBM's contracts with NASA over the years, less information was written about Federal Systems' involvement and technical contributions. Mainly because FSD was IBM, and reporters tended to lump together the two, although Federal Systems performed a much different role than providing standard computer hardware to NASA. Also, press release copy by both NASA and IBM tended to be truncated many times in interests of media column space. Further, FSD's technical work was "behind the scenes," difficult to explain and somewhat unexciting to report.

PROJECT MERCURY
Tracking & Ground Instrumentation

FSD's earliest contract was as a team member with Western Electric, Burns and Roe Inc., the Bendix Corporation and Bell Telephone Laboratories to develop the worldwide system of tracking and communications for Project Mercury awarded in July, 1959. FSD scientists developed the orbital mathematics required for tracking the Mercury spacecraft on its three-orbit mission. The following year the contract was extended. FSD then assumed responsibility for the mathematics (computer logic) involved in tracking the vehicle during the launch phase. Under a separate contract, IBM produced and installed two IBM 7090 computers at the Goddard Center, Greenbelt, Maryland.

One of the major entities of FSD's Space Systems Department was the organization devoted exclusively to the Mercury contract. Its responsibility was the coordination, planning, operations and control of IBM's entire support to NASA by the several centers in FSD, divisions of IBM and subcontractors. This total effort began with approximately 200 people, 130 of which were IBM employees located at Bethesda, Kingston, and Poughkeepsie, New York, Bermuda, Goddard, Cape Canaveral and Miami. This highly talented group was skilled in such disciplines as programming, mathematics, astronomy, computer engineering and design and communications.

FSD's responsibilities were to install and maintain these three large-scale computers (two 7090's at Goddard and one 709 in Bermuda); develop the mathematics and logical formation to enable the computer programs to supply required display and digital information; translate this formulation into a highly integrated set of computer programs, and to develop the special equipment enabling Cape Canaveral and Mercury Control Center complex to directly communicate with the Goddard and Bermuda computers as well as outlying radar sites. Computers were used throughout Mercury missions to process observations in the launch, orbit and re-entry phases and to supply a continuous up-to-date record of the non-environmental status of the spacecraft. The   Mercury   range   network consisted of eighteen radar and telemetry sites throughout the world. Mercury was a real-time system: speed and reliability in computing and in data flow were essential.

Through May 1962 ten project Mercury launches were successfully supported by the Launch Monitor Subsystem:

			Description	Occupant	Date
Mercury	Redstone	1	ballistic	unmanned	11/21/1960
Mercury	Redstone	2	ballistic	primate	1/31/1961
Mercury	Atlas 2		ballistic	unmanned	2/21/1961
Mercury	Redstone	BD	ballistic	unmanned	3/24/1961
Mercury	Atlas 3		orbital	unmanned	4/25/1961
Mercury	Redstone	3	ballistic	manned	5/5/1961
Mercury	Redstone	4	ballistic	manned	7/21/1961
Mercury	Atlas 4		orbital	unmanned	9/13/1961
Mercury	Atlas 6		3 orbits	manned	2/20/1962
Mercury	Atlas 7		3 orbits	manned	5/24/1962

New frontiers of space exploration and travel were driving pursuit of science and engineering as never-before experienced in history. The competition raised by the USSR further inspired and accelerated action in this area, creating large demand for scientists and engineers. The establishment of space objectives by Presidents Eisenhower and Kennedy further accelerated technology development and growth. It is only now, in hindsight of 50 years history that one understands the daring technology that was exploited with Project Mercury.

Discussions with one FSDer, Bob Campenni, who transferred into IBM's Federal Systems Division in 1962 from an IBM computer development organization based in Poughkeepsie, New York, revealed the following summary of his Project Mercury, Gemini and Apollo experiences from 1962 through 1968.

In 1962, NASA had just completed the first U.S. manned orbital flights that were dependent on a complex inter-connected system of hardware, software and people deployed on a worldwide basis, including floating and bobbing tracking ships in the oceans. This combination of technology and people was far from standard "off-the-shelf" components and skills. When one looked at the schematics of the interconnected systems, one's mind would glaze over with incredulity. It was difficult to comprehend this masterpiece and a daunting challenge to any newcomer to the project. Nonetheless, the results they produced were outstanding and extremely beneficial to all of humanity. Along the way, President Kennedy raised the bar with the manned lunar landing target for the decade of the 1960s. Not only was this objective achieved successfully, but it left the legacy of what the term "team" defines.

Only through the unselfish participation of thousands of individuals representing commercial and government organizations and banding together in great cooperation did success result. Bob went on to spend over 50 years in the computer industry and counts his Mercury/Gemini/Apollo experience as the highlight of a personal career and a hallmark of what can be achieved on the "national level" when a national problem has been identified and resolution is required. It is only necessary for the leaders to agree to pursue the challenge as a national objective and solicit the team approach embodied by NASA and encouraged by the U.S. leadership and its citizens.

Bob Campenni continues, in light of the above, FSD and Project Mercury should be put in more perspective. This will allow an understanding of how far we have come along the technology curve and to appreciate many of the concepts that exist today. First we must acknowledge the primitive nature of real-time processing and data communications in those early days. The concept of remote terminals was merely a thought in the late 1950s and punched cards were still the basic input/output medium. Operating system software was basic with very little multi-tasking capability. Data communications was virtually non-existent other than in "teletype" mode and high-speed data was a figment of the imagination. With that background, here are some of the areas that Federal Systems contributed.

In the hardware area there was a need to collect remote data in both low-speed and high-speed formats from around the world. This was accomplished with a special 7281 Data Communications Controller that multiplexed 32 subchannels of input/output data to and from teletype circuits, high-speed (1000 bits per second) receivers and transmitters, plot boards, strip chart recorders, wall maps, consoles, etc. Data communications was primarily teletype oriented with the 18 world-wide tracking stations and 1000-bps high-speed transmission between Cape Canaveral, Florida, and the Goddard Space Flight Center (GSFC) in Greenbelt, Maryland. Data communication between Bermuda and Cape Canaveral was non-existent in the early days and therefore necessitated an IBM 709 computer system in Bermuda to perform backup processing for Cape Canaveral. A cable was later laid between Bermuda and Florida to allow data communications at 1000 bps. Much of the other input/output hardware was provided through FSD contract and subcontract vendors.

In the area of system software, FSD and NASA programmers created a Real-Time Monitor and Real-Time Processors to perform the special programming required for Mission Support. In similar fashion, simulation software was developed to allow testing of the Real-Time Mission Programming. Given the complexity of this world-wide system, these programming efforts were far from trivial at the time. However, they have resulted in a legacy of transaction processors that pervade our society today (*e.g.*, Internet shopping, ATM transactions, on-line banking, travel reservations, eBay, etc.).

Any system, particularly a system as complex as the Mercury Tracking and Ground Communications Network must be thoroughly tested under conditions as close to actual operating conditions as possible. It is not enough to know that units and subsystems function properly; it must be established that all elements function together as part of an integrated subsystem or as a complete system. Thus, in addition to equipment unit testing and launch subsystem testing responsibilities, FSD was given responsibility for checking out of all computer-related elements of the total Mercury system. Called CADFISS (Computation and Data Flow Integrated Subsystem) testing, this world-wide network test was employed in Mercury (also Gemini, Apollo) launch countdowns to determine final tracking radar system readiness. Throughout the entire mission, CADFISS was also used to test remote-site readiness to continue supporting the mission. Thus CADFISS provided the means of evaluating a real-time ground support system in real time.

All the above reflects the precursing activity Federal Systems brought to the commercial marketplace. Today's generation benefits immensely from FSD's effort but has little awareness of the "gap" traveled in both the computer and communications industries unless these trivial facts end up in Wikipedia and Google search results.

PROJECTS GEMINI & APOLLO

Ground Based Data Processing — NASA Houston Real-Time Computer Complex (RTCC) and FSD Owego's On-Board Guidance Computers.

A difference in mission separated FSD's Washington Center space flight programs from its Owego Space Guidance Center programs. SGC emphasized fabrication of hardware for on-board in-flight control. By contrast, WSC concentrated on implementing ground-based flight information systems, which usually entailed the integration of IBM equipment with other devices.

The RTCC serves as a data processing and computing center. Raw data passes from the integrated mission control center to the RTCC which converts the data to useful information and makes it available to mission controllers. Approximately 200 on-site FSD personnel supported the complex. Under a later contract with NASA, the Mercury support service included analysis and programming which adapted the Mercury system to portions of the new Gemini program.

On-Board Guidance Computers

In 1962, FSD Owego began work on a guidance computer that would help steer the two-man Gemini capsule. This contract was worth $36 million and resulted in about 1,500 man-years over a four-year period to build and maintain a computer system that was 15 times more powerful than the IBM computer complex used to support Project Mercury. This system handled over 25 billion calculations a day during the time when the Gemini capsules were in flight. The computer weighed only 59 pounds and occupied only 1.35 cubic feet of space. In 1966, for the first time, the Gemini spacecraft was automatically guided through reentry by this on-board computer system.

By the end of the Gemini program in 1966, FSD's guidance computer aided in the accomplishment of these space firsts:
** First maneuvers by an orbiting spacecraft
** First rendezvous in space
** First docking space
** First navigation in space
** First rendezvous in initial orbit
** First on-board computer controlled reentry

IBM/FSD personnel supported each of the Apollo missions from Apollo 11's first landing on the moon in 1969 through the final missions of Apollo, Apollo 16 and 17, in 1972.

NASA awarded IBM/FSD a contract to support the Apollo/Soyuz joint U.S.-Soviet space venture scheduled in

1975 as well as contracts to provide computers, displays and programs for NASA's Space Shuttle scheduled for operation in the 1980s. This successful Apollo-Soyuz mission concluded NASA's Apollo series of space flights in 1975. For the next year, the Enterprise, NASA's first vehicle in America's Space Shuttle Program, would make its debut at Palmdale, California, carrying flight controllers and special hardware built by Federal Systems engineers.

SKYLAB

IBM supported the fourth manned space program of NASA. Skylab was designed for long-duration missions. It had two objectives: to prove that humans could live and work in space for extended periods and to expand our knowledge of solar astronomy well beyond Earth-based observations. In addition to the computer processing of orbital trajectory and spacecraft monitoring that served the previous manned flights of Mercury, Gemini and Apollo, Federal Systems Space Guidance Center, Owego, developed and produced special equipment for Skylab. Two on-board computers controlled the orientation of the laboratory throughout the mission. These computers, arranged redundantly for added reliability, were models of Owego's 4Pi computers. (Please refer to Chapter 9).

SPACE SHUTTLE

While IBM computers played key roles in each Space Shuttle flight, FSD provided the on-board computers under contract to the Space Division of Rockwell International Corporation. The computers were part of an Advanced System 4Pi avionics computer services. Input/Output Processors acted as interface between the computers and other orbiter systems.

SATURN

Saturn, the U.S.'s most powerful rocket system, was

designed to boost huge payloads into Earth orbit. The rocket has the capability of boosting over 45,000 lbs, transporting men to and from the moon and placing instruments on Mars and Venus. FSD Owego produced the digital computer to guide the Saturn booster. SGC engineers also conducted several studies in flight control.

Saturn V Instrument Unit (IU)

IBM established its Marshall Space Flight facility group in Huntsville, Alabama, to produce the instrument ring for the Saturn rocket for the Apollo moon landings program. IBM/FSD provided one entire section. It physically interconnected the third stage of the launch vehicle with the uppermost section containing the Apollo spacecraft. FSD was responsible for total systems integration, assembly and checkout. The IU involved a massive variety of unique interfaces. Saturn became the workhorse of the space age.

JPL – Jet Propulsion Laboratory
Space Flight Operations Facility

Under a contract with NASA, the Cal Tech Laboratory selected FSD to conduct a study of data-handling requirements. JPL had constructed a Space Flight Operations Facility (SFOF) as a command-control center to test spacecraft operations and to direct interplanetary exploration by unmanned probes such as Ranger, Mariner and Surveyor. This center became operational in 1964 with a great deal of IBM standard computer equipment installed.

The 1980s produced a bonanza of space firsts and FSD's shadow force of engineers and scientists continued to support NASA's exciting exploratory missions. The Soviets were also quite busy and continued to expand their space program.

Soviet cosmonauts returned to Earth after a record 185 days aboard the Salyut Space Station in 1980, and the U.S.

Voyager 1 space probe sent back pictures of Saturn, its moons and rings. Six new moons were discovered.

In 1981 the first U.S. space shuttle *Columbia* made its first maiden flight and NASA announced that two unusual meteorites found in the Antarctic may have originated from Mars. Also, in 1983 the U.S. space shuttle *Challenger* was launched on its maiden flight. In 1985 *Discovery* was launched as America's first exclusively military space mission, while at the same time the space shuttle *Atlantis* made its maiden flight.

In 1986 the space shuttle *Challenger* exploded on takeoff, killing all seven crew members. Voyager 2 flew by Uranus and discovered 10 more moons. In 1987 the U.S.S.R. launched its new heavy-lift rocket *Energiya* and Soviet cosmonaut Yuri Romanenmo returned to Earth after a record-setting 326 days in space aboard the Mir Space Station.

Almost two years later, in 1988, America launched the Space Shuttle *Discovery*, the first U.S. manned space mission since the *Challenger* disaster some 20 months earlier. As the decade of the 1980s came to a close, the Soviets launched their 2,000th Cosmos, satellite and NASA launched the *Galileo* space probe to Jupiter.

IBM and FSD continued to support NASA's many missions well into the early 1990s until the division was sold. FSDer's can look back with great pride on the work they accomplished in support of the Nation's space efforts for almost four decades. Many continue on this proud tradition as employees of Lockheed Martin.

Everything is simpler than you think, and more complex than you can imagine.
- Goethe

7. CONTROLLING THE SKIES
Air Traffic Control

The Air Commerce Act of 1926 was the cornerstone of the Federal government's regulation of civil aviation. Today the Federal Aviation Administration (FAA) is an agency of the Department of Transportation (DOT) with the authority to regulate and oversee all aspects of civil aviation in the United States.

With the approaching era of jet travel and a series of mid-air collisions, Congress passed the Federal Aviation Act of 1958. This was the same year that NASA was established following Russia's launching of the first artificial satellite. In 1967 the new DOT combined Federal responsibilities for air and surface transport and changed the name to the Federal Aviation Administration.

IBM had been engaged in studying air traffic control problems since 1955. From this continuous program a group of systems engineers was formed. This group combined operational experience in air traffic control with extensive digital computer and real-time systems background. One of the first steps taken was the hiring of controllers with operational experience. Another was using consultants from the areas of aviation not represented by IBM/FSD personnel.

A number of jobs were performed under contract with FAA. One consisted of a simulation program written to simulate the functions of an Air Route Traffic Control Center. Another was the flight generation program designed to generate flight-plan samples on the FAA 7090. In addition, FSD's group conducted a complete study and evaluation of all elements of both Air Defense and Air Traffic Control Systems in a Super Combat Center Deployment

Environment. In general the intent of this contract was to perform preliminary design work to aid FAA in the fullest use of the SAGE system for air traffic control.

By the mid-1970s, the FAA had achieved a semi-automated air traffic control system using both radar and computer technology, portions of which were supplied by IBM. In 1985, Federal Systems was awarded a major contract valued at more than $197 million to upgrade the computer systems supporting the 20 air traffic control centers in the United States. Two IBM 3083 BXl systems were to be installed at each center as a major step to modernize the air traffic control system.

When the FAA awarded the contract of the Advanced Automated System (AAS) it was recognized by FSD that an entirely new physical plant was needed due to the scope of the project. A site was selected in a newly constructed campus in Rockville Maryland, not far from FSD headquarters. Following IBM's financial problems of the early 1990s due to a variety of large-scale integration and systems requirements issues, the sale of Federal Systems to Loral caused the FAA program to run into severe technical difficulties. The program was cancelled and restarted twice, first in 1994 and later in 1996. The program was eventually cancelled at Lockheed Martin's request, parts of which were spun off into much more successful Display Screen Replacement (DSR) and HOST Replacement programs in 1998 and 1999 under Lockheed Martin's Air Traffic Management organization. Lockheed Martin also built the New En-Route Centre in 1998 and ATC centers in Taiwan, Argentina and South Korea.

Today, an inter-agency task force of experts from NASA and the Departments of Defense and Transportation are working on a dramatic transformation of air traffic control, called "NextGen," tripling the systems' capacities without tripling the workforce. NextGen does this by harnessing precise information from GPS satellites, advanced communications and sophisticated automation. Air traffic then could be managed from anywhere in the country, not just at major sites as is being done today. Given past history it remains to be seen whether such a radical program will survive the congress.

The transfer of know-how and experience in common cause is a noble endeavor.

\- Anonymous

8. SPREADING THE WORD FSD
Non-Federal & International Business Contracts

Since its formation, Federal Systems developed a strong capability to address and provide solutions to complex governmental problems. Beginning in early 1968, the many countries of IBM's World Trade Corporation (WTC) began using FSD resources to satisfy similar special commercial requirements worldwide. Their customer demands had increased rapidly, and WTC desired strong and fast response to these market conditions. To properly control and effectively coordinate FSD's response and to facilitate adequate and proper resource allocation, Federal Systems established its International Programs Office (IPO) under the direction of William J. Almon, an ex-West Pointer and excellent sales manager. Further, IBM domestic commercial divisions sought FSD's expertise in supplying special products, technical services and/or people to increasing demands in the U.S. private sector. These too were mostly handled by the International Group that produced the proposals, selected the personnel and handled project management duties where needed. Bill Almon's staff of specialists soon got used to his management style; he gave general orders and expected the staff to fill in the details and perform without much supervision. While doing this, at times he appeared and looked completely relaxed, but his eyes were fully engaged. Bill's talents soon came to the attention of IBM's senior management and he was selected as the IBM President's Chief of Staff. Almon later went on to a Vice Presidency running one of IBM's San Jose, California, facilities.

FSD teams were sent overseas to develop and implement

solutions to customer requirements in banking, newspapers, utilities, automobile manufacturing, railroad and television communications. During this same period, the New York Stock Exchange needed specialized help; the New York City Police Department required the development and installation of its SPRINT network, a command and control system for police cars, and the New York Power Pool at Guiderland, N.Y., needed technical support in developing its control center. FSD's expertise was put to use in the U.S. Postal Service's mechanization program.

International Banking

In the late 1960s, International Banks were faced with challenges that pre-cursed U.S. activity. Many countries such as the United Kingdom, France, Denmark, Sweden, Holland, Switzerland, Japan and Canada had large numbers of branch offices as part of their national banking systems, while in the U.S., banking legislation in the 1930's had restricted the potential for nationwide banking. This difference led to opportunities for FSD talent to be assigned to address the unique challenge of national banking. FSD's experience with Real-Time Operating Systems in the NASA world uniquely positioned FSD to address these challenges. Two banks in the UK, Westminster and Barclays, were in need of such assistance. Westminster had recently obtained government approval to merge with another large national bank (National Provincial Bank) and become the National Westminster Bank (NatWest). In similar fashion, Barclays Bank was pursuing a merger with Martins Bank that would end up with the Big 3 UK banks being National Westminster, Barclays and Lloyds. The challenge at hand was to develop and install on-line branch office transaction accounting systems. Neither banks' thousands of branches (in the city of London alone) could process their city-wide customer demands for teller check cashing and currency disposal. A bank customer could not cash his check or draw funds from his own bank unless he used the branch in which he originally established his account. This was an incredible bottleneck to business development.

There was another factor driving IBM UK to pursue this outside assistance from Federal Systems. Both Westminster and Barclays had non-IBM mainframes and posed a serious competitive challenge for the future terminal networks. In response, FSD assembled two five-person teams that were assigned to support both Westminster and Barclays Banks. Personnel assigned were selected from FSD's major locations in Washington, Houston and Los Angeles.

The Barclays' effort entailed designing the on-line Branch Banking System in the greater London area. Unfortunately, the envisioned system did not come to fruition because the UK government did not approve the merger with Martins Bank for monopolistic reasons. FSD personnel were deployed to other challenges, such as a study of the Foreign Exchange operations, leading to improved processes. This activity not only benefited Barclays but for one FSD assignee who had banking accounts with Barclays in both London and Los Angeles, it provided some humorous interplay with the local branch manager over circumventing the serious overdraft policies of UK banking. For the two years the FSD assignee was stationed in London, he regularly waited for the branch manager to call and apologize for their internal inefficiencies showing his account going into "overdraft" status. This became the signal to transfer additional funds from Los Angeles.

On the NatWest side, FSD assistance was applied to the design of an on-line transaction processing system using new-technology "smart" workstations. Another major effort was mounted to perform project management and programming assistance related to UK "Decimalization" by 1970. This challenge was heightened by the merger of Westminster and National Provincial Banks into National

Westminster Bank with a new "Cheque Clearing" System. In addition, during the late 1960s FSD personnel also provided technical support to the Swiss Bank Corporation through various studies on future Information Technology direction.

In discussing banking support on an international basis, it should be noted that the prime driver behind this international support was FSD President Bob O. Evans. Bob was an outstanding executive and a brilliant engineer with a fabulous ability to remember details and names. He led IBM into the System 360, a major initiative that revolutionized the computer industry. He became President in 1964 and later moved on to President of IBM's Systems Development Division. He exploited the system integration capabilities of FSD into the non-federal world, such as Bay Area Regional Transportation Automatic Fare Equipment, Japanese newspaper publishing, Japanese broadcasting, Strathcona Refinery Process Control and others.

There is even a story of a famous telephone call to FSD's Rockville Headquarters from Wehrner von Braun asking to "speak with the chief engineer" (which was Bob Evans). That call was integral to FSD's involvement with the Apollo Saturn program. When Bob moved on to SDD, he also moved key personnel to SDD to migrate the "systems capability" into IBM's commercial product line. It was not surprising that Bob would also take IBM into the satellite area (Satellite Business Systems, managed by a long time ex-FSD Vice President, Phil Whittaker).

It is also not surprising that in the mid-1970s IBM executives re-organized the corporation, and the "systems capability" house of SDD morphed into a narrower Systems Communication Division. In hindsight, this reorganization was the beginning of IBM's decline in the late 1970s through the early 1990s as a credible "systems house."

In France, the Credit Lyonaise Bank in Paris and the Royal Bank of Denmark also required similar assistance and FSD system specialists were assigned for varying periods of time. The French Railroad, , required an FSD specialist for assistance in a managment information system.

Japan, Germany & Sweden

In Japan, FSD teams working with IBM Japan technicians successfully developed and implemented a revolutionary real-time newspaper computerized publishing system for the major Japanese newspaper, *Asahi Shimbun*. Also, FSD personnel participated in developing a system of television control for NHK, the major Japanese television network.

FSD engineers developed a Film-Reader Recorder System for the German government Postal Telephone and Telegraph Service (PTT) for utility billing of its customers.

In Sweden, an FSD Systems Engineer/management specialist spent two years assisting Volvo in developing its internal management information and processing system for automobile production.

For decades FSD expertise was employed in the United States on domestic state-wide and city-wide technical projects because the various IBM product divisions needed assistance. In New York City, for instance, FSD developed and implemented the SPRINT system for Police Dispatch. This project was upgraded on several occasions and as late as 2002, IBM was contracting with the Office of Management Analysis and Planning to study and develop a disaster recovery plan for SPRINT operations.

Federal Systems personnel were involved in implementing a Special Operations System for the New York Stock Exchange. As late as 1988, FSD personnel were assisting the Federal Aviation Administration on en-route air traffic control and Post Office mechanization.

These domestic and international projects brought significant revenues to FSD, learning experiences in diverse business enterprises, and special hardware product sales. The sale of one large IBM System/360 Model 65 to Barclays

Bank is one example of the enormous influence FSD work had on standard IBM Product sales throughout the world. For example: NASA required 5 IBM 7094 II's for the Real-time Computer Complex (RTCC) at Mission Control, Houston to support the Mercury, Gemini, and Apollo moon flights; and upgrades to IBM 360/Model 75's over the years to the current Space Shuttle flights which began in the late 1970s. NASA also received an IBM System/360 Model 91 in 1968 (Please refer to Chapter 6 - NASA, for a complete review of IBM's and FSD's support of the U.S. space effort).

While it is recognized that IBM salesmen sold thousands of computers to the U.S. government over the years, FSD personnel due to their strengths in solving complex problems, contributed greatly to standard IBM computer product sales and revenues.

For example:

2 IBM 7094 Space Flight Operations Facility (SFOF) JPL.
2 IBM 7040 " "
3 IBM 1401 " "
1   IBM 7094 White Sands Missile Range
1   IBM 7044 " "
1   IBM 7044/1401 Space Detection & Tracking System
1   IBM 7090 Strategic Air Command AIDS (SAC)
1   IBM 1401/1410 Electro-Magnetic Intelligence USAF 473L
2 IBM 7090 NASA Bermuda
2 IBM 7090 USAF 465L System
2 IBM 7090 BEMEWS 474L System
1   IBM 7090 USN Target Acquisition System
1   IBM 1410 Defense Communications Agency
1   IBM 7090 FSD Owego operations Computer Center

In November, 1993, one month before its sale to the Loral Corporation, Federal Systems landed a major contract with the Internal Revenue Service to design and implement a system to scan completed IRS forms, subtract the images, leaving only information. This information would be converted to data and stored electronically. This contract totaled $1.3 billion and was to extend for 15 years.

In addition, in January, 1994, the Loral Corporation was awarded a $155 million contract by the United States Postal Service to manufacture and install over 3,000 automated mail sorters, based on Federal Systems expertise and know-how. Further, the Justice Department, the Internal Revenue Service and the DOD Sustaining Base Information Services contracts had the future revenue potential of more than $5 billion.

Thus, the question will always remain: Was selling the Federal Systems Company a good idea? While it provided an immediate influx of cash, what were its long-term costs to today's IBM?

Management Systems

FSD's contracts to help both civil government agencies and the military manage their business activities encompassed systems operational support, systems integration and test and complete systems development for various government agencies. A sampling would include:

1. Bluestreak — Navy Cost Information System
2. Office of Economic Opportunity Information Handling
3. Data Processing for National Meteorological Service
4. Master Forces & Routing Methods System for the U.S. Army
5. STINFO — Information Handling System for Army Research Office
6. Data Handling for the National Institutes of Health
7. Data Evaluation System for the Puerto Rican School System.

SABRE (Semi-Automatic Business Related Environment), was American Airlines' (AMR) on-line, interactive reservations system. FSD systems engineering personnel designed and developed the system. SABRE was one of the first examples of on-line interactive systems. This system generated over \$15 million in profits for American Airlines in only its first year of operation. Needless to say, travel agents and the general public loved it. It was a bold move by AMR to invest almost \$40 million in such a computer-based system — the first online real-time system.

The Information Technology (IT) industry learns from current solutions and develops newer ones. FSD had a role in helping to create the Internet. With its pioneering work on the SAGE network for the USAF; databases, both internal and on the Internet, with specialized databases created for the CIA and the NSA; imagery, now common with the photo mapping applications it created for NASA and NOA. In effect, FSD worked on the "bleeding edge" of IT, with its work products commercialized years later by others. Indeed, it was IBM's "Silent Service."

Federal Systems, then, has come full circle. From its beginning as Military Products during World War II to its end as the Federal Systems Company in 1993, its scientists and engineers, its managers and support personnel, can look back proudly on the work it performed for our Nation.

It should have come as no surprise when the organization was sold to Loral. Lou Gerstner's history with RJR Nabisco telegraphed his intentions. During his first six months as CEO at RJR he sold off \$3.5 billion of assets. Hiring Jerry York to take the meat-ax to IBM's airplanes, real estate, people, and facilities, doomed Federal Systems, since neither York nor Gerstner knew much about computers or FSD's highly sophisticated and experienced workforce . . . a workforce that could have redirected much of its talent to addressing fee-based consulting on a worldwide basis. This chapter has presented the myriad projects and programs FSD performed in the private and non-military government sectors. This more than proves the point that selling FSD was a mistake.

A lot of people lost their jobs at IBM, some at FSD, as well. IBM reduced its global workforce from 405,000 in the late 1980s to a low of 215,000 in the early 1990s. Needless to say, this almost 50% reduction in active payroll involved losing many talented people. For the most part, however, FSD was spared the wholesale firings that IBM itself endured, since significant numbers of its personnel who were working under contracts simply transferred to its new owner, Loral Corporation, and then on to Lockheed Martin.

It was a magnificent run. An era had ended.

A modern computer hovers between the obsolescent and the non-existent.
-Sydney Brenner

9. WHAT'S REAL-TIME?
Structured Programming - Iterative & Incremental
Software Development & Real Time Applications

In early 1974, the U.S. Air Force's Rome Air Development Laboratory (RADC) sponsored a series of efforts to define and evaluate what was becoming known as Modern Programming Practices. This effort was a major positive influence on the structured programming revolution which resulted in a whole new generation of programmers learning new techniques for specifying, designing, implementing, testing and maintaining computer software. One of the early advocates was Dr. Harlan Mills.

Structured programming was seen as an exciting idea among software developers. RADC awarded Federal Systems' Dr. Mills a contract in 1974/75 to define structured programming with a detailed set of guidelines as to how to implement structured programming. The U.S. Army's Computer Systems Command amended the contract and provided co-sponsorship to this contract. The result of Dr. Mills' work was a fifteen-volume report. It wasn't until late 1979, however, that a textbook presentation of Dr. Mills' ideas was available.

Mr. Richard C. Linger, a Senior Programmer in FSD, created and taught the first technical and management course in structured programming in FSD and is co-author with Dr. Mills and Mr. Bernard Witt of *Structured Programming: Theory and Practice*, Addison-Wesley Publishing Company, Inc.

Unstructured programming was used for decades and is a procedural program. Statements are executed in sequence as written. Structured programming came into being and was made possible by high-level languages and packages of related programs that could be easily flow charted with software diagramming techniques understandable to non-programmers.

Iterative & Incremental Software Development (IID)

Dr. Mills' well-known *Top Down Programming in Large Systems*, written in the early 1970s, promoted iterative development, and he was one of its earliest proponents. The FSD Trident Submarine project beginning in 1972 was the first documented FSD application of IID. The Trident project was a high visibility, life-critical system containing more than one million lines of code and became the command and control system for the first Trident submarine. This system had to be completed and delivered on time or FSD would face late penalties of $100,000 per day.

Another early FSD project involving IID was the U.S. Navy's LAMPS (Light Airborne Multipurpose System) described briefly in Chapter 5, Defending America's Skies. LAMPS was a four-year, 200 man-year effort involving millions of lines of code. Again, both of these large programming projects were delivered on time. Project Mercury, it has been suggested, was the seed bed out of which grew the Federal Systems Division, and the history of incremental software development. During the late 1970s and on into the 1980s FSD applied IID for the NASA Space Shuttle program.

Real-Time Systems for Federal Applications

In September 1981 two Federal Systems senior engineers, Messrs. P.F. Olsen and R.J. Orrange of FSD's Owego facility, authored a definitive overview of FSD's highly creative work in complex, on-line, real-time systems and their conventional data processing counterparts, in IBM's Journal of Research & Development (Vol. 25, Number 5, dated 9/81). This report discussed the highly specialized FSD-developed computers used in both military and NASA space programs over many decades. It is reprinted here in its entirety. Copyright permission appears on page 118.

P. F. Olsen R. J. Orrange

Real-Time Systems for Federal Applications: A Review of Significant Technological Developments

The Federal Systems Division of IBM has been heavily involved in complex, on-line, real-time systems for over twenty-five years. In this paper a representative sample of these programs are reviewed, and an evaluation of the significant lessons learned from this wealth of experience is presented. The key issues which differentiate real-time systems from their more conventional data processing counterparts are identified and their implications are discussed. This leads to some conclusions regarding the kind of commitment that is necessary in order to succeed in the area of real-time applications.

Introduction

The basic mission of the Federal Systems Division of IBM (FSD) is to serve the special data and signal processing needs of the federal government. Some of these needs can be satisfied with standard products, but other requirements can only be met with a combination of hardware, software, and services configured to satisfy the user's applications.

Fulfillment of FSD's mission has involved a broad range of challenging applications and technology, yet there is a common thread among these widely varied programs: FSD has consistently been involved in on-line, real-time applications. Real-time data processing is characterized by highly structured and repetitive tasks, high speed, stability considerations for closed-loop processing, automated fault detection and decision making, and customized input and output. The following sections of this paper describe programs that trace the evolutionary progress of FSD in real-time systems. An overview of representative FSD systems is given and the FSD design environment is addressed. This is followed by sections on technological implications, real-time digital processing technology, reliability and availability, and total responsibility. The last section of the paper describes the LAMPS (Light Airborne Multi-Purpose System) program to illustrate the nature of one aspect of the division's current activity.

Twenty-five years of challenge:
An overview of representative FSD systems

FSD was launched in 1955 as the Military Products Division of IBM. At that time it had two major contracts: the data processing centers for the SAGE (Semi-Automatic Ground Environment) air defense system, and the onboard AN/ASQ-38 navigation and bombing system for the B-52 aircraft. These programs involved real-time systems that were significant technological advances over any-thing preceding them, and they were representative of the great majority of FSD programs ever since. For example, SAGE was the first of many complex *ground-based command and control systems* for FSD. Subsequent FSD programs in this category include:

** SABRE (Semi-Automatic Business-Related Environment) — American Airlines' on-line interactive reservations system [1].

** IBM 9020 — the multiprocessing computer system that is the core of the FAA's National Airspace System for coordinated air-traffic control throughout the United States [2, 3].

** RTCC — the space program's Real-Time Computer Complex in Houston, Texas, discussed further in an accompanying article in this issue [4]. FSD has had an important and growing role in every manned space program: Mercury, Gemini, Apollo, Skylab, and currently, Space Shuttle.

** LPS — Shuttle Launch Processing System [4].

** LCC — Saturn Launch Control Complex [4],

** DFCS — the Drone Formation Control System, at White Sands Missile Range, which allows control from the ground of aircraft flying in close formation and/or performing maneuvers that would not be feasible in piloted aircraft [5].

** SPRINT — a command and control system for police cars in New York City.

** The control center for the New York Power Pool, Guiderland, NY.

** GPS—the control segment for the Global Positioning System, a recent (1980) ground-based control system awarded to FSD. This system features precise position measurement of the constellation of satellites through continuous worldwide monitoring. The satellites relay the ground-based computations to suitably equipped users, permitting precise navigation worldwide in all weather. IBM is prime contractor to the Air Force and will provide the entire system, including RF, data processing, and software elements. Similarly, the B-52 avionics system of the 1930s was the first of many *onboard data processing systems* for aircraft, missiles, satellites, ships, and submarines—the kind of products and systems that have become a way of life for FSD. A sampling of subsequent programs in this category would include

** Titan II (ASC-15 computer)—IBM's first onboard computer in the missile and space field. It was a drum machine, and was also used in Saturn I and IB.

** —the Orbiting Astronomical Observatory on-board processor, which utilized quad-redundant components [6].

** Gemini—manned orbiting vehicle computer.

** Saturn V—Apollo launch vehicle computer.

** Space Shuttle—redundant computer.

** System/4 Pi—FSD's family of avionics computers. Thousands of 4 Pi machines have been delivered on scores of different applications, including over 2000 missile guidance units for the Harpoon program, more than 1400 avionics computers for the A-7 aircraft, and additional computers for the Fill, FBI 11, A-6, and EA6B aircraft.

** AN/BQQ-5—the Navy's first digital submarine sonar system: an integrated onboard signal processing and control complex. The technology of the time was such that 600 System/360 Model 65s would be required to duplicate the function of its digital beam-former—a special-purpose processor designed in 1970 [7],

** APR-38 Wild Weasel—an airborne radar signal-pro cessing system consisting of 14 separate units and more than 50 antennas which detect, locate, and classify radar emitters.

** MCS—The Modular Computer Series, the latest generation of System/4 Pi machines. By an interesting coincidence, the first several hundred MCS processors will be used on the B-52 aircraft as part of an overall replacement of avionics equipment produced in the 1950s and 1960s [8].

** E-3A AWACS—FSD is responsible for the data processing system in two generations of the Airborne Warning And Control System which flies aboard a modified Boeing 707 aircraft [9].

Finally, the combination of the SAGE and B-52 was a precursor of yet a third area of FSD business: *system integration responsibility for onboard command and control systems*—for submarines, ships, aircraft, and space vehicles. This has become a premier business of FSD and has included such major programs as the following:

** Saturn V Instrument Unit (IU) — Saturn V was the launch vehicle (rocket) for the Apollo moon landing program, and IBM provided one entire section of it. The IU physically interconnected the third stage of the three-gigagram launch vehicle with the top section containing the Apollo spacecraft. It served as the central control and communication element of Saturn V. IBM was responsible for total system integration, assembly, and checkout of the more than sixty pieces of equipment contained in the IU. The Guidance Computer, Data Adapter (I/O Unit), Switch Selector (signal routing unit), Operational Flight Software, and Ground Command and Control Software were designed and provided by FSD [10].

** TRIDENT CCS—the TRIDENT Command and Control System consists of five subsystems requiring the integration of more than a million lines of software and over 100 different major hardware units [11].

** LAMPS—FSD has total responsibility for the overall performance of an electronics-packed, ship-based helicopter plus all of the associated shipboard equipment [12], The LAMPS program involves all of the issues discussed in this paper and we will return to LAMPS subsequently to examine the total range of responsibility and commitment required to succeed in the world of real-time systems.

The foregoing comprise a representative sample of the scores of real-time applications that FSD has addressed over the past quarter century. A further examination of the environment in which these systems operate, and the conditions under which they are procured, will help to highlight some of the unique issues that must be faced in this area.

Application-driven design environment

FSD systems usually have to be tailored to the specific environment of each application. "Onboard" systems, for example, usually fit into very limited (and sometimes very odd-shaped) volumes. They must be lightweight and low-power and must operate reliably over ambient temperatures ranging from -55°C to 71*C, while subjected to such hostile environments as shock, vibration, dust, and salt spray.

Often the most significant accommodations that must be made are the unconventional interfaces inherent in many real-time applications. For example, the basic inputs of a sonar system are audio signals. Airborne radar and electronic countermeasures systems operate at the other end of the spectrum with carrier frequencies in the gigahertz range. These systems detect, analyze, and emit various types of electromagnetic pulses—with each type having its own unique waveforms, pulse repetition rates, carrier frequencies, and other specialized characteristics.

Typical avionic and missile data processing equipment must interact with a wide range of sensor and control subsystems, each with its own unique data and control interfaces. For example, in determining present position, an avionic system may utilize inputs from the gyros, accelerometers, and gimbal angle sensors of an inertia! sensor system; position fixes from radar or visual sights; Doppler velocity measurements and altitude readings from radar or barometric subsystems; data from active ground- or satellite-based navigation reference systems; plus manual inputs which select modes of operation, identify fixes, etc.

On the output side, an avionic computer may provide commands and data to align, stabilize, and correct an inertial platform from which it obtains velocity and altitude measurements. The overall operation of the radar system is usually computer-controlled, and most FSD systems include computer-driven dynamic situation displays. Ground-based systems, such as SAGE, SABRE, the FAA En-route Traffic Control System, and various military command and control systems, require special communications networks to interconnect data processing equipment at widely dispersed sites [1, 3].

One current example of such a real-time communications processing system is SACDIN (Strategic Air Command Digital Network)—designed to provide reliable, secure communications among the geographically dispersed facilities of the Strategic Air Command. The three headquarters commands are linked by SACDIN subnet communications processors to one another and to 16 dispersed base communications processors (BCPs). SACDIN further interconnects the BCPs with 20 aircraft wing command posts and 26 missile base command posts which, in turn, are interconnected to 128 launch control centers.

As an element of the worldwide military command and control system, SACDIN must accommodate a wide variety of message protocols and must interface with a number of different communications systems. SACDIN involves such functions as message organization/process-ing/accountability/journaling; recovery and reconfiguration upon loss of equipment or interconnectivity; encryption/decryption and overall security of both message traffic and data base; diagnostic capability and status reporting; and data base management and system traffic control [13].

A major challenge in this class of real-time communication systems is to simultaneously provide two somewhat mutually exclusive capabilities:

1. The levels of security—including partitioning, access control, and accountability, as well as encryption— that are a fundamental system requirement in a strategic command and control network.

2. The certainty of access to/by qualified users throughout the network — even in the face of massive losses of equipment and communication links — that is equally essential to the function of a strategic command and control network.

The Saturn V IU program illustrates the range of different I/O requirements that may be encountered in a single real-time application. With more than sixty different pieces of equipment to integrate, the IU involved a massive variety of unique interfaces. A single unit, the IBM-built Launch Vehicle Data Adapter (LVDA), provided an interface for 402 different signals [10].

It is evident that one of the singular and significant realities of real-time systems is the extent to which the user's requirements drive the design. This characteristic has required FSD to be involved in a broad range of technical disciplines that stretch beyond the boundaries of conventional signal and data processing requirements.

Technological implications

FSD's pursuit of real-time systems has involved a wide range of technical disciplines: radar, stellar mechanics aircraft and missile flight dynamics, RF antennas, sonar, space navigation, ballistics, inertia! alignment, phased-array beam-forming, precision tracking of high-speed aircraft, and the integration and operational debugging of complex on-line, real-time systems. FSD has developed a strong capability in many forms of advanced signal processing, including sonar, precision emitter location systems, emitter classification and identification, synthetic aperture radar, moving-target detection, image processing, Kalman filtering, and seismic data processing [14-17].

Some of this expertise has been put to good use by the rest of the Corporation. A recent example is the IBM 3838 Array Processor, which will be discussed in more detail subsequently. In addition, the major challenges which FSD has faced in real-time control software development have led to substantial contributions to software engineering within IBM [18].

The trend toward standard digital intra-system interfaces has kept FSD at the forefront of emerging technology in fiber optics. Work was complete in 1980 on a Navy contract to develop and demonstrate both a 1-MHz, serial, MIL-STD-1353B-compatible fiber optic data bus [19], and a 50-MHz fiber optic data bus suitable for the forecasted avionic requirements for the 1990 time period [20]. In addition, PSD is actively participating in the drafting of a new military standard for fiber optic data buses [21].

A unique technological ramification of FSD's operational environment is a continuing reliance on and significant progress in magnetic core, drum, and disk storage technology. Many applications include requirements for non-volatility (instant recovery after interruption of power) that can only be met by core main store and magnetic media mass storage [22].

The capacity of a SAGE drum was 0.4 megabits. The FSD drum used on numerous programs since 1970 has a 15-megabit capacity in a much smaller volume and meets the full military specifications (MIL-Spec) range of shock, vibration, temperature, etc. Each 4096-word SAGE core memory required approximately 2.8 m* to house the memory with its electronics [23].

The full MIL-Spec core memory in FSD's Modular Computer Series provides 64 000 (32-bit) words in 0.004 $m\backslash$ including electronics — an improvement of more than 10 000:1 in packaging density. And, of course, performance, reliability, and power consumption have also improved considerably. In fact, FSD core memory technology development continues to cut the volume, failure rate, and power dissipation of core memories in half every few years. During the last decade, DOD (Department of Defense) programs were generally followers rather than leaders in the development of mainstream data processing technology, such as integrated circuits. The sheer volume of commercial applications made it far more practical to "hitch a ride" by hardening commercially available technology rather than trying to amortize high development and tooling costs over much smaller DOD procurement quantities.

The advent of VLSI (very large scale integration) is forcing a change in this pattern. Heretofore it has been quite practical to design unique processors using readily available unit logic and the modest levels of integration available to date. But the higher the level of integration, the more specific the function tends to be—and thus the less readily adaptable to multiple applications. At the same time, the characteristics of VLSI are just as critical in opening up new DOD applications as in accomplishing similar objectives for commercial and consumer applications. In fact, the cost, power, reliability, weight, and volume advantages of VLSI are absolutely essential to the practicability and effectiveness in satisfying many future DOD requirements.

The government has recognized this situation and has allocated substantial funding for basic VLSI technology development under its VHSIC (very high speed integrated circuits) and DARPA (Defense Advanced Research Projects Agency) VLSI programs [24]. FSD was given one of the nine contracts awarded to industry for VHSIC Phase Zero government sponsorship. But even before the VHSIC program, the division had recognized the need for a strong VLSI capability and had focused the requisite facilities and skills in FSD Manassas [23]. The objective of FSD's VLSI advanced development program is compatible with that of the VHSIC program: development of submicron technology during the 1980s. In the process, FSD expects to make a significant contribution to IBM's total capability in this key technological area.

Real-time digital processing technology

The problems addressed by computer technology can be characterized by two generic problem types [26]:

** General-purpose processing, somewhat unstructured in nature, and

** Signal processing, which by comparison is highly structured.

FSD develops computers to address both of these generic problem types. There has been a remarkable evolution in these computers through the 1970s as both component technology and organizational technology have experienced continuing improvement. A review of that evolutionary process should aid perception of two of FSD's more recent development efforts:

** The IBM Advanced System/4 Pi CC-2 general-purpose computer, and

** The IBM 3838 Array Processor.

Table 1 lists a number of significant computers developed by the division in recent years. If one were to characterize these members of the IBM System/4 Pi family of avionic computers by performance, four categories could be formed:

** Low-performance (SP-0, ML-0) general-purpose processors, generally embedded in a subsystem such as complex I/O control/multiplexor or displays—emphasizing low power, volume, and weight rather than performance as the key parameters.

Table 1   Representative FSD-developed computers.

Model	Date of prototype delivery	Type of machine	Thousands of operations per second	Circuit technology (delay time)	Degree of MSI	Control components (cycle time)	Main store (access time)
AP-1	1971	GPC	360	Standard TTL (11 ns)	Very Low	Hardwired SSI	8K × 18 Core (1 μs)
SP-0	1971	GPC	230	Standard TTL (11 ns)	Very Low	Hardwired SSI	4K × 18 Core (1.3 μs)
CC-1	1973	GPC	740	Schottky TTL (6 ns) introduced	Low	256 × 4 PROM (80 ns)	8K × 18 Core (1 μs)
AP-101	1973	GPC	550	Schottky TTL	Low-Medium	256 × 4 PROM (80 ns)	8K × 18 Core (0.9 μs)
ASP	1975	SP	to 60,000	Schottky & Low Power Schottky TTL (11 ns)	High	1K × 1 RAM (70 ns)	4K × 36 FSU (0.8 μs)
ML-1	1975	GPC	550	Dutchess LSI Custom	(LSI)	512 × 4 PROM (70 ns)	16K × 8 Core (0.85 μs)
ARP	1976	SP	to 20,000	Schottky & Low Power Schottky TTL	High	512 × 4 PROM (60 ns)	1K × 36 RAM (0.07 μs)
ML-0	1977	GPC	550	Schottky & Low Power Schottky TTL	High	512 × 4 PROM (60 ns)	16K × 36 RAM (0.85 μs)
IBM 3838	1977	SP	to 30,000*	Schottky TTL	High-Very High	1K × 1 RAM (70 ns)	8K × 36 FSU (0.8 μs)
AP-101C	1978	GPC	600	Schottky TTL	Very High	2K × 8 PROM (90 ns)	32K × 18 Core (0.8 μs)
CC-2	1980	GPC	2000	Super Schottky TTL (3 ns)	Very High	1K × 1 RAM (40 ns)	32K × 18 Core (0.8 μs)
Series/1 (MIL)	1980	GPC	350	Dutchess LSI Custom	(LSI)	512 × 8 PROM (85 ns)	64K × 18 DRAM (0.66 μs)

Legend: GPC – general-purpose computer.
SP – signal processor.
* – floating-point operations.
First-level package = circuit carrier, typically flatpack.
Third-level package = card interconnection, typically multi-layer backplane.
Dutchess is IBM's name for a 100-TTL-circuit-per-chip technology.

MSI—Medium-Scale Integration.
LSI—Large-Scale Integration.
TTL—Transistor-Transistor Logic.
DRAM—Dynamic Random Access Memory.
FSU—Functional Storage Unit (semiconductor memory).

** Medium-performance [AP-1, AP-101, ML-1, AP-101C, Series/1 (MIL)] general-purpose processors, typically used as the central computer of an avionics system. The price/performance of these machines has tracked the improvements experienced with general-purpose machines (Fig. 1).

** High-performance (CC-1, CC-2) general-purpose processors used in command/control systems. In addition to the obvious benefits of increased speed and density in both logic and memories, these machines have also taken advantage of advanced organization techniques to provide higher levels of performance.

Signal processors (ASP, ARP, IBM 3838), whose performance is several orders of magnitude greater than general-purpose processors as a result of the nature of the signal processing problem and the willingness to spend hardware to achieve very high performance. The contrast between high-performance general-purpose processing and signal processing will be described in greater detail subsequently.

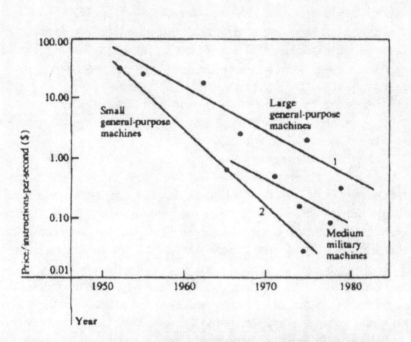

Figure 1 Computer price/performance trends. Curve 1 represents an average improvement for large general-purpose machines of 15% per year; Curve 2 represents an average improvement for small general-purpose machines of 25% per year.

The advance of technology through the 1970s can be seen in the data presented in Table 1. The technologies listed are those employed in the prototype hardware. Frequently the storage technologies, both control store and main store, have been retrofitted in production.

** *Logic technology* -- Logic technologies have seen marked improvement in performance and, more importantly, in circuit density. Schottky devices were introduced early in the 1970s and offered a doubling of performance but at higher power dissipation. The need to constrain power in many military and space applications motivated the introduction of low-power Schottky devices. More recently, advances in circuit geometry have permitted continued improvement in the circuit speed-power product, particularly the introduction of high-performance Super Schottky.

** *Memory technology* -- Neither of the two elements of memory technology, main memory and control memory, have seen performance improvements to match those of logic technology. Both have, however, seen dramatic improvement in density, as discussed earlier. Most DOD real-time applications have required non-volatile memory. However, the cost, performance, and volume advantages of monolithic memory have become so compelling that DOD users are exploring such means as battery-backup power, memory hierarchies, and fault-tolerant memory organization to take advantage of these benefits.

The increased usage of microprogram control has been a result of several factors. Foremost was the availability of a cost-effective technology beginning in 1972 with reliable programmable read-only memory (PROM). Although the integration level of PROMs has evolved from the original 256 x 4-bit chips to the current 2K x 8-bit chips (K - 1024), the performance of these devices has held fairly constant, probably due to the dimensional constraints of the programmable cells. Those seeking higher performance for microprogram control have used static RAM, which offers comparable density and improved performance but unfortunately is volatile and requires supporting IMPL (initial microprogram load) logic.

The advantages of microprogram control have typically been ascribed to the flexibility it offers in implementing custom functions. Of greater importance to the computer designer is the regularity afforded by microprogram control. Control logic (as opposed to data flow logic) tends to be irregular, complex, unsuitable to increased integration levels, and subject to high probability of engineering change. The attractiveness of microprogram control regularity resulted not only in increasingly horizontal micro-words (i.e., using many bits to directly control data flow elements, rather than imposing combinational logic between an information-dense microword and the data flow elements), but also in the use of PROMs and PLAs (programmable logic arrays) wherever possible to replace combinational logic.

** *Packaging technology* -- The increasing circuit integration levels (MSI, LSI, PROMs, RAMs, PLAs) and the desire to pack more function into available volume produced new problems in component cooling and interconnection. About 160 integrated circuits on two multi-layer boards sharing a frame and a 196-pin connector could be mounted on a 4 Pi page or card. Heat was conducted from the page frame to the cooling plenum through a pair of ears with limited cross section. The maximum power dissipation allowable (to keep junction temperatures below 125°C with conductive cooling) was about 20 W. As power density increased, several approaches to cooling evolved, including integral heat exchangers, air flow through the page frame, larger thermal surface contact area, and, when possible, direct air impingement on the components. The current IBM Advanced System/4 Pi Modular Computer Series (MCS) page permits indirect cooling of up to 40 W by employing a pair of wedge-locked surfaces.

The technology evolution just described has culminated in a pair of products which exemplify the merging of organization and technology. Both the IBM 3838 and the CC-2 designs illustrate the impact of technology on organizational issues and the organizational approaches to coping with the limitation of technology.

** *Signal processing* -- Perhaps the most challenging and significant work done by FSD in special digital processors has been in applications involving signal processing. The signal processing problem typically involves repetitive, highly structured computations on very large data sets. To do this in real time requires very high throughput—tens *of* millions of floating-point operations per second in some cases. To provide this level of performance under the size, weight, power restrictions, and operating environment encountered in military applications is extremely challenging.

This challenge has been successfully met by a combination of tactics. First of all, the highly structured nature of signal processing tasks has allowed generality to be traded off for optimized performance. The AN/BQQ-5 sonar signal processors designed in 1970 were basically hardwired fast-Fourier-transform (FFT) array processors whose input and output data structures were fixed. Subsequent FSD products, such as the Advanced Signal Processor (ASP) [27], used on several different military contracts, and the IBM 3838

Array Processor [17], have been programmable processors designed for a range of signal processing applications.

The Advanced Signal Processor is an IBM product developed primarily for sonar applications and has been designated the DOD standard signal processor, AN/UYS-1. Its sonar applications include operation in long-range patrol aircraft, fleet protection helicopters and destroyers, and in submarines. Other applications include ground data processing of Air Force and NASA satellite data. The IBM 3838 Array Processor is a floating-point commercial derivative of the ASP. A wide range of signal processing algorithms are microprogrammed into its highly efficient and pipelined arithmetic processors, including recursive and finite impulse response filters, fast Fourier transforms, envelope magnitude detection, and quadratic interpolation.

These designs also reflect the fact that many of the operations being performed are independent from one another—such as an element-by-element vector sum. This allows both pipelined computation to increase the throughput of a given processing element and parallel processing elements to achieve the desired total processing capacity.

In other applications, advanced statistical processing techniques have been used with general-purpose computers *to overcome* limitations in input signal quality due to overt energy interference, operating range, practical constraints on the size and sensitivity of airborne or space-borne sensors, and phenomena with inherently poor signal-to-noise ratios. One of many examples that could be cited is the work in Kalman filtering done in conjunction with the Safeguard Anti-Ballistic Missile (ABM) program. Techniques were developed by IBM to greatly improve tracking capability with a given processing capacity through the use of extended Kalman filters with adaptive tuning and chi-squared validation techniques using corrected covariance. These developments were supported by IBM's own tools and methodologies for filter design and system simulation evaluation [28].

Reliability and availability

The generally accepted definition of reliability is the probability that the system will perform satisfactorily for a specified period of time when operated within its specified environmental conditions

[29]. In the early 1950s, when FSD entered into defense business, it was revolutionary to see this performance parameter specified.

When it became intuitively obvious that single equipments could not satisfy performance demands, massive redundancy of equipment was utilized. The SAGE system required continuous operation around the clock to monitor airspace coverage; loss of the system for a failure and subsequent repair time was intolerable. Therefore, two systems capable of performing the same function were placed on line, thus negating the problems associated with a single failure. The small probability of loss of two systems at the same' time was addressed through overlapping coverage from adjacent sites. Since there were few constraints applied on the design regarding weight, volume, and power, the system redundancy solution was acceptable.

In the avionics world, these constraints became more meaningful. White the airborne mission was much shorter in time and non-continuous in nature, the technology of the day (vacuum tubes) combined with the complexity of the system designs raised grave concerns about the ability to satisfactorily complete a mission. Initially, these concerns were addressed by redundancy in the deployment of aircraft.

In the mid-1950s, R. Mettler performed a study for DOD in which he quantified the probabilistic nature of the reliability problem and focused attention on this parameter in future system designs. A joint government-industry advisory group on reliability of electronic equipment (AGREE) was formed. The work pioneered by this committee led to the evolution of reliability from an art to the technical discipline known today.

The decade of the 1960s brought FSD into the missile and space age. For missile guidance hardware, the most critical design parameters were light weight, low power, and computation accuracy. Required computational speeds did not demand state-of-the-art breakthroughs; however, the environmental considerations did demand solid state componentry (in that era, germanium devices with proven reliability).

The advent of manned spaceflight led to the most stringent reliability requirements ever placed on computer hardware, i.e., a probability of success of 0.99 for a mission duration of 250 hours [30]. To satisfy this requirement, a simplex computer would require a mean-time-between-system-failure of 25,000 hours, which was not achievable with state-of-the-art technology. A unique design employing triple modular redundancy with intermediate "voting" was developed for Saturn V which permitted individual piece-part failures to occur without affecting the correct system output. This design was packaged in a single lightweight frame fabricated from magnesium-lithium, and incorporated internal fluid cooling to maintain semiconductor junction temperatures less than 40°C for reliable device operations. By packaging the redundant computer into a single unit, significant savings in weight and volume were achieved.

Another significant reliability achievement in space applications was the design of the processor and the large memory for the Orbiting Astronomical Observatory [31]. This unmanned vehicle was required to operate for one year in orbit. A unique redundancy design was employed at the circuit level, using a quad-component arrangement which tolerated shorts or opens in individual components. The systems were launched and operated error-free for four years until shut down by ground command.

In the late 1960s and early 1970s, significant technology changes were being incorporated into hardware design. Logic circuitry with about five gates per chip was common in 196S. During the 1970s, MSI (medium scale integration) ranging from 30 to 100 gates per chip was employed in avionics computers. By the mid-1980s, logic densities of 20,000 gates per chip will be applied in military computers, with a corresponding improvement in per-gate failure rates. The higher-density circuits have also permitted fault-detection circuitry covering 99% of computer faults to be incorporated within the critical weight, power, and volume restrictions associated with many military applications.

But despite tremendous improvements in reliability, failures will still occur. And many real-time systems involve processes too critical to allow a failure to disrupt normal operation. The Space Shuttle onboard flight-control system provides an excellent example. It controls the spacecraft through all mission phases, including re-entry and landing [32]. Once again, redundancy is a necessity.

The Space Shuttle flight control system must satisfy a NASA requirement of fail operational/fail safe (FO/FS). This means that there is to be no impact on system performance in the event of a single failure and that the system must provide a safe return even after two failures.

A total of five identical computers are used. Four of these run identical programs and are fully synchronized. By comparing outputs, a failed computer can readily be detected, isolated, and eliminated from the system. The remaining three computers can then cope with the possibility of yet a second failure by comparing outputs in the same manner. The fifth machine runs a different backup program which protects against the possibility of a generic failure in the identical software used by the four primary computers.

Comparison of outputs among the four primary machines is performed by the computers themselves, with each of them passing its output data to one other computer for comparison. Under the assumption of non-simultaneous hardware failures, agreement between any two computers provides assurance that both are current. When there is a failure, the pattern of agreement/disagreement among comparison pairs easily isolates the faulty machine.

Synchronization is maintained on a software, rather than hardware, basis. Each of the four primary computers informs the others when it is ready to begin the next task in the common sequence they are all executing. No computer proceeds with a task until the others are also ready to proceed with it.

On Space Shuttle, all data communication between system elements occurs via 24 independent data buses and each computer has access to all 24 buses. For flight-critical data, these are used in redundant sets of four buses with each of the four controlled by a different computer. When one computer intends to read input data, it pre-notifies the other three, then commands the input. All four computers thus receive identical, simultaneous inputs from each sensor, eliminating the need for any correction or interpolation.

In addition, all critical sensors are provided in redundant sets of three, each controlled from a different computer. Since all buses can operate simultaneously, and the start of each task is synchronized, all three sensors are read at the same time. This arrangement allows for detection, isolation, and elimination of a failure anywhere in the system: computer, data bus, or sensor.

Total responsibility

Much of FSD's business consists of delivery of products (hardware, software, subsystems) to the government or another contractor. The involvement in other programs is different: As SPC (system prime contractor), FSD is responsible to the government for all aspects of system performance, from development to deployment and field operations, throughout the life of the program. SPC responsibilities consist of many or all of the following:

** Total system performance—including design, development, and delivery and field support of the system.

** Specification of interfaces between associate contractors.

** Design, development, and procurement of avionics/electronics and supporting software.

** Integration in the laboratory

** Testing and evaluation of the total system in the field.

** Specification of reliability and maintainability system requirements and assurance that they are met.

** Development of integrated logistic support (ILS) requirements and establishment of a program to ensure support in the field.

** Definition of training requirements and development of a training program, including acquisition and turnkey operation of training devices.

** Cost and technical performance tracking and reporting.

** Control and accountability for total system configuration status (change control of hardware and software).

Any major system must be developed through an iterative process of analysis, development, integration, and testing culminating in a complete set of validated specifications and plans for the production phase.

IBM's objectives in the integration of a major system, possibly encompassing avionics, air vehicles, shipboard electronics, and support facilities, plus a variety of technical and management disciplines and interfaces, is to ensure that the system:

** meets the required operational objectives;

** satisfies the design-to-cost boundary conditions and provides minimum life-cycle costs, including the indirect costs of manning and logistics;

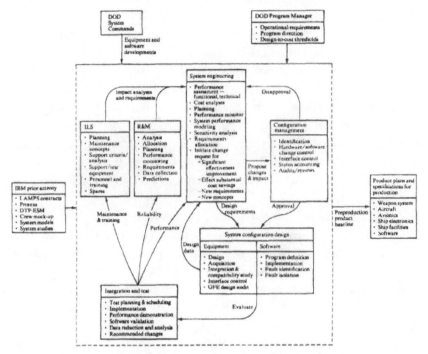

Figure 2 Role of the system prime contractor. ILS—Integrated Logistics Support; R&M—Reliability and Maintainability; DTP—Design To Price; ESM—Electronic System Management; GFE—Government Furnished Equipment.

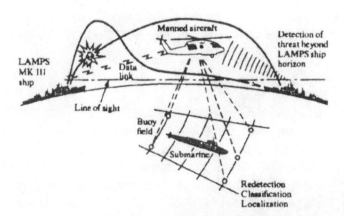

Figure 3 LAMPS MK III Weapon System—mission concept.

**minimizes risk relative to performance requirements, predicted availability, cost thresholds, and schedule milestones;

** provides visibility to the government and utilizes related government expertise and developments.

Attainment of these objectives is a significant challenge which can only be satisfied by establishing and executing, on a timely basis, the proper system engineering plan and methodology which allows 1) definition of the complete requirements for each of the subsystems and interfaces with proper consideration to constraints of cost and risk; and 2) early integration, test, and evaluation to either verify performance or identify problems in time for corrective action to be initiated.

In the development phase the SPC performs the following role: implementation of the development test program; documentation and planning for the production programs; management coordination and technical support for the government program manager; and support of government test programs.

The interrelationship between the tasks required to successfully design/develop/integrate and test the LAMPS weapon system is illustrated in Fig. 2. In essence, the SPC must act in the interest of the DOD agency in all matters in order to ensure orderly progression of the system development program through the major phased-procurement milestones [Defense System Acquisition Review Council (DSARC)] leading to the introduction of a new mission capability into the DOD inventory. The LAMPS program provides the best example of the technological complexity and the types of activities involved when FSD is called on to accept total responsibility for a major military undertaking.

LAMPS program

Early in 1974 FSD was selected as the SPC for the Navy's LAMPS MK III program. FSD is responsible to the Navy for the total performance of the LAMPS MK III weapon system, which embodies the integration of the parent ship (frigate, destroyer, cruiser) and the manned aircraft (SH-60B Seahawk helicopter) operating from that ship for both Anti-Submarine Warfare (ASW) and Anti-Ship Surveillance and Targeting (ASST) missions (Fig. 3).

The LAMPS in weapon system performs the four classical

phases of the ASW problem (detection, localization, classification, and attack) in an interlocking manner, in which the ship does the initial detection and classification, and its aircraft accomplishes redetection, localization, reclassification, and attack. Information sharing, by means of a wide-band duplex data link between the parent ship and the helicopter, commences when the helicopter is launched. The data link is synchronized, and navigation is initialized relative to the ship position when the helicopter is first airborne and prior to transit to the threat area. As sonobuoys are deployed, the acoustic data are transmitted along with radar or electronic warfare information by data link to the ship for processing. These data are also processed on board the helicopter. Command and control, though maintained by the ship, can be delegated to the helicopter after contact is established and the target is being tracked by the airborne unit. Final localization and attack are accomplished independently by the helicopter using its total acoustic and non-acoustic sensor onboard capability.

In the ASST mission, the LAMPS aircraft provides a mobile elevated platform for observing, identifying, and localizing threat platforms beyond the parent ship's horizon. Primary sensors used in the ASST mission are radar and electronic warfare support measures equipment. Thus, the effective surveillance, detection, and targeting ranges of the parent ship are greatly extended to permit targeting of platforms that might launch a missile attack. The helicopter performs ASST at significant distances from the threat platform to minimize its vulnerability to attack.

System integration requirements and challenges

The LAMPS MK III weapon system posed system integration challenges of a degree and complexity that mandated primary emphasis on this design parameter during development. The problem was compounded by having to check out new computer programs in parallel with the introduction of new data processing and display equipment and newly denned interfaces, both in air and ship systems. Both IBM and the Navy recognized the importance of utilizing formal documentation to ensure successful integration both within the LAMPS system and with major ship systems with which LAMPS must interface.

The LAMPS ship and air system is comprised of eight

computers and 140 black boxes weighing approximately 900 kg on the helicopter and 3200 kg on board the ship. A total of almost 600 000 words of operational and maintenance software is resident in eight different computers. An additional 600 000 words of simulation software was required. The major subsystems in LAMPS MK III include the following:

** Navigation—Teledyne AN/APM-217 Doppler radar set, Collins AWARN-118 TACAN, and Texas Instruments AN/APS-124 radar set.

** Communications—two Collins AN/ARC-159 UHF radio sets for line-of-sight, and one Collins AN/ARC-174 HF radio set for over-the-horizon.

** Acoustics — Hughes AN/UYQ-21 shipboard acoustic display, and two EDMAC Corp. AN/ARR-75 radio receiving sets for sonobuoy signals.

** Magnetic Detection—Texas Instruments AN/ASQ-81 magnetic detecting set including a towed magnetometer.

** Data Processing — two IBM Proteus AN/UYS-1 processors to process acoustic data aboard the aircraft and the ship, two Control Data Corporation AN/AYK-K standard airborne computers, and two Univac AN.UYK-20 standard shipboard computers. In addition, one Univac AN/UYK-7 standard shipboard computer is required in the Combat Direction System (CDS).

The data link, which moves data simultaneously in both directions, consists of a Sierra Research Corporation AN/ARQ-44 radio set in the aircraft with two directional antennas and Sierra's AN/SRQ-44 set on board ship with a high-gain directional antenna. The data carried include clear or secure voice, secure computer, radar/IFF or acoustic data, and sonobuoy command tones.

In order to facilitate the software development and integration of hardware and software, FSD set up a facility which includes a System/370 computer to provide tactical conditions for land-based test operations and to simulate sensors and interfaces prior to operational hardware delivery. This facility has subsequently been identified as the "LAMPS Land-Based Test Site" by the Navy. A sizeable number of Navy personnel are resident, both for training purposes and also to validate the operational and technical testing being performed. Other facets of the Land-Based Test Site facility include:

** Sonar signal processing and display laboratory.

** Avionics subsystem integration laboratory, where prime hardware from 25 contractors is integrated and tested. It is configured so that eight sensor subsystems, which go on the helicopter, can be independently tested or integrated into one system.

** Shipboard laboratory, which is a simulated combat-information center containing four operator consoles and their associated electronics, which are used to evaluate operator functions and to validate shipboard software, and also can be interconnected with the avionics bench or a helicopter in flight.

** Air system master bench, a full-scale mockup of the Seahawk's cabin area. It has been used to train Navy crews, to integrate mission avionics hardware and software, and to test performance.

** Hangar, landing pad, and control tower, which are used to house two Seahawk helicopters for avionics installation and checkout and functional flight test operations. A realistic antisubmarine warfare mission can be flown from Owego, New York, using the ship laboratory both for functional evaluation and crew training.

The first flight took place in December 1979 [33] and the combined ship/air weapon system performance demonstration took place during the last three weeks in February 1980. The third week was used by the Navy to try to "break" the system; it could not be broken. Since then, a total-weapon-system test has been completed and testing by the Navy has begun. Throughout the history of this project as well as others mentioned in this paper, IBM has striven to fulfill the role of a responsible system developer.

References:

1. David R. Jarema and Edward H. Sussenguth, "IBM Data Communications: A Quarter Century of Evolution and Progress," *IBMJ. Res. Develop.* 25, 391-404 (1981, this issue).

2. "An Application-Oriented Multiprocessing System," *IBM Syst. J.* 6, No. 2 (1967).

3. D. L. Walker, "Getting It Together . . . For Safety in the Skies," *Technical Directions* (IBM Federal Systems Division) 1, 2-9 (1975); and "Flexibility, Reliability . . . An Extra Margin of Safety in the Skies,"

Technical Directions, 16-25 (1976).

4. S. E. James, "Evolution of Real-Time Computer Systems for Manned Spaceflight." *IBM J. Res. Develop.* 25,417-428 (1981, this issue).

5. K. D. Rehm and S. P. Remza, "Drone Formation Control System," *Technical Directions* 5, 3-13 (1979).

6. A. E. Cooper and W. T. Chow, "Development of Onboard Space Computer Systems," *IBMJ. Res. Develop.* 20, 5-19 (1976).

7. "A Big Step in ASW Systems," *Technical Directions 1*,2-5 (1975).

8. R. L. Carberry, "The Modular Computer Series — Spanning Military Requirements into the 21 st Century," *Technical Directions* 5, 2-5 (1979).

9. "E-3A: USAF's Serial Sentry," *Technical Directions* 2,2-8 (1976).

10. *Astrionics System Handbook,* National Aeronautic!) and Space Administration, Huntsville, AL, August 196S.

11. C. Manarin, "FSD People and Their Part in Building the Trident Submarine," *FSD Magazine* 3, 2-7 (1979).

12. W. M. Gaddes and R. J. Orrange, "LAMPS (Light Airborne Multi-Purpose System)," *Technical Directions* 4, 3-15 (1978).

13. C. W. Sturgeon and Z. G. Tygielski, "SAC Digital Network," *Technicul Directions* 5, 16-22 (1979).

14. R. E. Blahut. "On Digital Processing of Coherent Radio Signals," *Technical Directions* 3, 20-30 (1977).

15. R. Bernstein, "Digital Image Processing of Earth Observation Sensor Data," *IBM J. Res. Develop.* 20, 40-57 (1976).

16. F. H. Schlee, C. 3. Standish, and N. F. Toda, "Divergence in the Kalman Filter." *AIAA Journal* 5, 1114-1120 (1967).

17. *IBM 3838 Array Processor,* Order No. G520-3134-1, avail able through IBM branch offices.

18. H. D. Mills, D. O'Neill. R. C. Linger, M. Dyer, and R. E. Quinnan, "The Management of Software Engineering," Parts I-V, *IBM Syst. J.* 19, 414-477 (1980).

19. R. Betts, "An Aircraft 1553B Compatible Fiber Optic Interconnect System," *Proceedings—Second Multiplex Data Bus Conference,* Air Force Systems Command, Wright-Patterson Air Force Base, Dayton, OH, 1978.

20. R. Betts and M. Swails, "High Speed Fiber Optic Data Bus for Distributed Processing Systems," *Proceedings of the Electro/80 Convention,* Electronic Conventions. El Segundo, CA, 1980.

21. Society of Automotive Engineers A2K Task Group, "Proposed Standard Fiber Optics Mechanization of an Aircraft Internal Time Division Command/Response Multiplex Data Bus," *Draft DOD-STD-XXFO,*

AFAUAAD-3, Wright-Patterson Air Force Base, Dayton, OH, 1979.

22. L. D. Stevens, "The Evolution of Magnetic Storage," *IBM J. Res. Develop.* 25, 663-675 (1981, this issue).

23. R. R. Everett, C. A. Zraket, and H. D. Benington, "SAGE — A Data-Processing System for Air Defense," *Proceedings of the Eastern Joint Computer Conference,* Washington, DC, 1957, pp. 148-155.

24. M. Sbobat, "Unprecedented Corporate Commitment Seen in VHSIC Program"; and M. Shohat and E. W. Martin, "View from the Top"; *Military Electronics/Countermeasures* 6, 26-36, 16-24 (1980).

25. J. Robertson, "IBM Federal Division Unveils Custom LSI Production Unit," *Electronics News,* May 19, 1980, p. 114; also "FSD Inaugurates Advanced Development Program Facilities at Manassas," *Manassas Focus,* May 12, 1980 Special Edition, IBM FSD Communications Department, Manassas, VA.

26. R. W. Bergeman, "Computer Technology," *Technical Directions* 6, 13-18 (1980).

27. Robert R. Kelly and Seymour Charton, "A Programmable Digital Signal Processor Evaluated for Radar Applications," *Technical Directions* 4, 16-20 (1978).

28. K. R. Brown, A. O. Cohen, E. F. Harrold, and G. W. Johnson, "Safeguard Track Filter Design — Part 1," *Final Report, Contract DASG60-75-C-WZ7,* IBM Federal Systems Division, Manassas, VA, 1975.

29. M. Y. Hsiao, W. C. Carter, J. W. Thomas, and W. R. Stringfellow, "Reliability, Availability, and Serviceability of IBM Computer Systems: A Quarter Century of Progress," *IBM J. Res. Develop.* 25, 453-465 (1981, this issue).

30. J. E. Anderson and F. J. Maori, "Multiple Redundancy Applications in a Computer," *Proceedings, Annual Reliability and Maintainability Symposium,* Washington, DC, Jan. 10-12, 1967, pp. 553-562.

31. J. E.Anderson, "7 Years of OAO," 1968 Product Assurance Conference and Technical Exhibit, Hofstra University, Hempstead, NY.

32. R. E. Poupard, "Space Shuttle Redundant Computer Operations," *Signal,* February 1979, pp. 41-44; see also J. R. Sklaroff, "Redundancy   Management   Technique   for   Space Shuttle

Computers," *IBMJ. Res. Develop.* 20,20-28 (1976).

33. David R. Griffiths, "LAMPS MK III Moves to Navy Testing," *Aviation Week <fc Space Technology,* June 30, 1980, pp. 47-48.

Received June 12, 1980; revised November 6, 1980

The authors are located at the IBM Federal Systems Division facility, Owego, New York 13827.

*Science is always wrong: it never solves a problem
without creating ten more.*
- Anonymous

10. INSIDE SCOOP
IRAD - Independent Research & Development

Federal Systems, since its inception has had an in-house research and development program whose objectve was to contribute to the advancement in those areas of military-oriented technology where IBM possessed an expert capability to fill a recognized military need.

The FSD program was grouped into six sub-programs:
** Computer Technology
** Communications Technology
** Space Technology
** Information Handling
** Lasers
** Marine Sciences

The IRAD program aimed at combining technical excellence and foreseeable usefulness. It was designed to strike a balance between applied research and applied development. FSD's IRAD program was independent from IBM's corporate research activity whose basic mission was to provide the scientific, engineering, and technological basis for future

IBM products and services: IBM's research activities have established and maintained an eminent position in the industrial and scientific world. Such research is legion and well known through its patent process.

Each FSD Center had its own funded IRAD tasks employing dozens of scientists and engineers guided by the understanding of the applied research activities of the Department of Defense and other agencies.

The Computer Technology Program was directly aimed at

improving the overall reliability and performance of special processing systems required by the military.

The Communications Program is the application of both existing and new digital processing techniques and theory to communications.

In the area of Space Technology the IRAD program focused on the spectrum of information technology/handling required for future space missions.

In the general field of information handling, major tasks were directed at image processing.

The laser program conducts research into new materials and laser concepts.

And finally, in the field of Marine Sciences, research was directed at automatic data acquisition and reduction techniques to broaden understanding of the whole area of oceanography and undersea warfare.

The IRAD program annually received joint funding from the Government and IBM. Results of the studies appear in final reports fully available to both government agencies and IBM personnel.

Looking backward over four decades, it can be credibly stated that FSD's huge investment in manpower for military research and development products, systems and technical services produced enormous dividends for both the company and the government. While FSD people were not directly on the battle line, as a shadow force behind the warriors, their contributions to our nation's defense and space exploits were nothing short of magnificent.

In closing out this Chapter, it can be said that we are a scientific civilization: that means a civilization in which knowledge and its integrity are crucial. Science, after all is only a word for knowledge and there is no absolute knowledge. All information is imperfect — it has to be treated with great humility.*

* J. Bronowski, *The Ascent Of Man*

Performance is the Best Strategy.
  - Anonymous

11. FSD'S HARDWARE STORE

Technical Specifications of FSD Hardware

All three centers designed, developed, fabricated, and delivered special hardware equipment to both NASA and the United States Armed Forces as well as to other defense contractors. While a great many projects were and still remain classified, the following data is not. The pages which follow represent a wide variety of FSD products -all developed under NASA and military contracts. FSD performed as a prime weapons systems contractor or as a sub-contractor to other defense contractors. The following is a sampling of major products.

INDEX

In addition to the above, FSD developed a wide variety of hardware products whose specifications are not available. Several of these were either prototypes, one of a kind or became modified versions that entered production under different designs in the same family of computers. See Chapter 9, "Real-Time Systems for Federal Applications: A Review of Significant Technological Developments" by P.F. Olsen and R.J. Orrange.

The rapid pace of technology has seen a remarkable evolution in logic, memory, packaging and reliability beginning in the late 1960s. By the mid-1980s and early 1990s, these changes were being incorporated into hardware designs, not just in the computer field but in everyday household electronics products. When first designed and developed the products shown were state-of-the-art. Today in the 21st century, they seem archaic for technology has advanced "light years" beyond these early developments.

AN/ASQ-38 WEAPON CONTROL SYSTEM, OFFENSIVE

The B-52 was developed by the Boeing Airplane Company under prime contract to the Air Force Systems Command, USAF. The IBM Federal Systems Division has over-all management coordination responsibility for the development and integration of the AN/ASQ-38 Weapon Control System. Under a prime contract to the AFSC, IBM produces the AN/ASB-9 A Bombing-Navigation System which, when combined with four ancillary systems, makes up the complete weapon system.

ELEMENTS:
AN/ASB-9A Bombing Navigation System
APN/89A Doppler Radar

MD-1 Automatic Astro-Compass
AN/AJN-8 Heading Vertical Reference System
KS-32A Radar Recording Camera

The AN/ASB-9A Bombing-Navigation System consists of 400-cycle analog systems for solving the navigation and bombing problems and radar data presentation equipment to display search radar video intelligence. The ASB-9A supplies stabilization information to the ancillary equipment and heading error signals to the aircraft's autopilot.

CHARACTERISTICS:

The ASB-9A Bomb/Navigation System calculates and compensates for such variables as air speed, altitude, drift, earth's rotation, bomb ballistics, roll, pitch, and yaw to guide the B-52 to a predetermined release point over the target — drops the bomb within a 1,000 foot radius of the target — photographs the damage and guides the aircraft back to the base.

The integrated system allows precision bombing from altitudes of more than 60,000 feet and all-weather bombing or ballistic missile runs of more than 1,000 miles. A low-level capability is made possible by an Advanced Capability Radar (ACR) System which literally looks ahead of the speeding airplane and shows the commander the exact profile of the approaching terrain many miles before he gets to it. He can then hedge-hop over the terrain at high speed. The ACR will also show an over-all view of the area so that the pilot can take advantage of valleys and low spots along his route for cover.

The elements of the AN/ASB-9A system comprise a Navigation Computer Group, Bombing Computer Group, Advanced Capability Radar, Radar Data Presentation Set (RDPS), and Pilots Display Equipment. The Navigation Computer is a combination analog and digital computer that solves both the long and short range navigation problems based on aircraft heading and velocity. It provides automatic flight on a great circle course to within 200 miles of destination — after which it switches to planar navigation. Present aircraft position in latitude and longitude is continuously computed.

Three ancillary navigation devices feed data into the ASB-9A navigation computer to correct its dead-reckoning navigation. The first is the MD-1 Astro Compass, an electromechanical, celestial navigation device that automatically provides a true-heading signal by referencing stars, the sun, or the planets. The second device is the AJA-8 Heading Vertical Reference System that converts magnetic heading indications from the N-1 Compass System into the true heading by correcting for magnetic variation and earth rotation. The third navigation device is the APN-89A Doppler Radar which performs automatic and continuous computation of ground speed and drift angle utilizing the Doppler effect. Doppler radars make use of the change in the observed frequency of a radar echo due to the relative motion between the target and radar. When the distance between target and radar is decreasing, the observed frequency is higher than the emitted frequency, and vice versa. This shift in frequency identifies moving targets.

Wind data information is acquired either by a High-Speed Bombing Radar (HSBR) or the Doppler Radar and stored in the navigational computer for continuous dead-reckoning navigation. Return-to-base heading can be computed from the most recent wind data stored in the computer. In the wind mode, the APN-89A Doppler Radar supplies ground speed and ground track angle to the ASB-9A Bomb Navigation System to compute accurate wind data for both bombing and navigation. In the memory mode the APN-89A receives ground speed data from the ASB-9A maintains the Doppler ground speed computation of a nearly correct value, and supplies a drift angle error signal to keep the APN-89A antenna aligned with the approximate ground track.

The Radar Data Presentation Set provides an electronic display of the area beneath the aircraft and contains the controls and circuits required to present the various cathode ray tube displays. The RDPS display is the Topographical Comparator which provides a means for map matching. A map, photograph, or predicted radar display is simultaneously projected on a 5-inch and a 10-inch cathode ray tube face. The picture on the small tube provides a means of target reference to the operator. Radar returns over the target area are superimposed over the projected image on the 10-inch tube face. When the match is perfect, the crosshairs are on target.

The Bombing Computer provides automatic flight control during the bombing run and automatic ballistic correction for bomb type, altitude and true air speed. It operates in search, track, wind determination, and synchronous bomb modes and provides instrumentation for airspeed and altitude indications. The Bombing Computer takes into consideration the bomb data, computes the correct time and position for releasing the bomb and guides the aircraft on the bombing run. Characteristics of the bomb being used can be manually set into the Ballistics Data Storage Unit.

Bomb-damage assessment and reconnaissance photographs are taken by the radar-recording camera from the rear of the radar scope. The camera can take PPI, Sector Scan, or direct radar exposures, as well as optical exposures.

FEATURES:

The system is designed in small modules for installation flexibility and ease of maintenance. The controls and indicators for the computer radar and presentation equipment were conveniently grouped in a console area for simplicity of operation.

Pluggable modules permit a building-block arrangement for system growth.

Built-in test points permit malfunctions to be isolated to a pluggable unit.

The computer analyzes and displays its own failure symptoms so that components can be replaced with on-board spares in flight.

Components have been tested for life endurance and the ability to withstand the extremes of the aircraft's operational environment.

All electronic units are packaged in cylindrical, hermetically sealed copper cans and cooled by forced air to limit maximum hot-spot temperature. All resistors are derated by a factor of three to one to four to one. Eighty-five percent of the vacuum tubes are operated at less than 50 per cent of rated plate dissipation.

Unit Test Equipment is used to check out all pluggable units on an individual basis on the ground. The entire system is subjected to the most rigorous environmental tests including:

High radio interference fields

30-G impact shocks

Sustained vibration

Salt-spray and fungus Sand and dust blasts

Temperatures up to 160°F at 95 percent humidity

AN/ASB-9A BOMBING NAVIGATION SYSTEM

The AN/ASB-9A BNS is a modification of the ASB-9 system. Development of the modifications were carried out at the IBM Owego facility under Air Force prime contracts.

DESCRIPTION:

The AN/ASB-9 A BNS, part of the AN/ASQ-38 BNS, contains analog computers for solving the navigation and bombing problems and presentation equipment for displaying radar video intelligence obtained from search radar equipment. The majority of the units in the system are relatively small, pluggable modules for installation flexibility and ease of maintenance. The controls and indicators for the Computer, Radar, and Presentation equipment are conveniently grouped in a console area for simplicity of operation. The console and other units of the system are interconnected with each other and with other associated systems by pluggable cables.

APPLICATIONS:

The AN/ASB-9A BNS is designed to automatically guide the aircraft to the release point, provide radar sighting, accurately drop the bomb on the target and direct the aircraft to the return destination. The system also provides radar search capability, takes radar presentation photographs, and provides missile launch equipment with certain basic information.

FEATURES:

Among the features which characterize the ASB-9 A BNS are reliability, flexibility and maintainability:

• Reliable operation is achieved by the use of certain design concepts, certified and tested components, and strong field support.
• Flexibility is built into the equipment by the use of pluggable modules, permitting a building block arrangement for system growth.
• Operational flexibility is also achieved by the use of redundant modes.
• Built-in test problems and test points facilitate the maintenance of the system by allowing malfunctions to be isolated readily to a pluggable unit.

CHARACTERISTICS:

The AN/ASB-9A is an electronic and electromechanical analog computer with search radar and radar presentation as part of the output equipment.

The number of units included in the major groups and their approximate weights are as follows:

Group	No. of Units	Total Weight
Bomb Computer	51	265 lbs
Navigation Computer	23	150
Advanced Capabilities Radar	15	450
RDPS	50	310
Pilots Display	24	150
Power Supplies	13	200
Miscellaneous	5	90

These figures do not include racks, cabling, or Government-furnished equipment.

The operational characteristics of the system are classified but can be obtained from the manuals covering the system.

All equipment not required in the AN/ASB-9A console for instrumentation and control purposes is located in remote areas in the aircraft. These non-indicating devices include timers, relay frames, integrators, calibrators, converters, selectors, transducers, generators and power supplies.

ELEMENTS:

The AN/ASB-9A BNS consists of a Navigation Computer Group, Bombing Computer Group, Advanced Capability Radar, Radar Data Presentation Set, Pilots Display Equipment, and associated power supplies.

The Main Digital Computer and the Emergency Digital Computer are discussed under "Airborne Data Processors". The Advanced Capability Radar and data displays of the AN/ASQ-38 System are found under "Airborne Data Presentation."

To provide additional strike capability for the B-52 the IBM SGC designed the tie-in of the system with the WS-131B (Hound Dog) missile and provided the equipment and engineering necessary for integrating the WS-138A (Skybolt) missile system with the AN/ASQ-38.

WS-131B (Hound-Dog)

The WS-131B missile tie-in with this system is now operational. The engineering effort involved the engineering design of the AN/ASQ-38 WSCO and the WS-131B missile system and thereby provided the critical data necessary to ensure missile strike accuracy.

The interconnection of AN/ASQ-38 and WS-131B involved providing, in analog form, those parameters required by the WS-131 digital conversion equipment. Signals and excitations generated with the AN/ASQ-38 BNS that provide position and velocity vector information are involved.

WS-138A (Skybolt)

R&D models of the analog to digital converter are being fabricated and tested, and flight tests are being conducted by the associated contractors for evaluation of the tie-in and the A-D conversion equipment.

This integration involved interconnecting the source points of several parameters required by the guidance system of the WS-138A missile system. All possible static and dynamic loading problems that would be imposed on the

AN/ASQ-38(V) system by the proposed tie-in were investigated for purposes of avoiding system degradation. The resultant interconnection between the AN/ASQ-38 (V) BNS and the WS-138A system was defined on system drawings for use by the airframe manufacturer.

The over-all tie-in involved the design of analog-to-digital conversion equipment necessary to supply the requirements of the missile system pre-launch computer. The signal conversion is accomplished through the use of electromechanical servos, encoders, miniaturized solid state amplifiers and an independent power supply. The conversion unit receives certain analog outputs of the AN/ASQ-38(V) WCSO and provides the missile with the same information in digital form.

REFERENCE DATA:
Information regarding the system may be found in various Air Force manuals, such as: T.O. llbl-ASB-4-21 HOI; T.O. 11 BIO-18-3-2 FMI Computer Group; and T.O. 11B31-^1-2-12 FMI Radar and Data Presentation Sets.

AN/ASQ-28(V) BOMBING NAVIGATION SUBSYSTEM

The AN/ASQ-28(V) BNS was limited to prototype development. A second prototype was built and delivered to North American Aviation for flight test in XB-70 No.3.

The AN/ASQ-28(V) Bombing-Navigation Subsystem was developed at the IBM Space Guidance Center for use in a supersonic, manned, weapon system. This equipment is a completely integrated subsystem, designed for Mach-3 environments.

CAPABILITIES:

The system can perform the entire bombing and navigation functions without restrictions of weather, time of day, or location of target. Integration with other subsystems of the weapon system is planned to enhance the performance of the BNS and provide maximum utilization of information in the BNS for other weapon system functions.

The basic system has growth potential to accept improved sensors such as a focused radar data processor, infrared detectors, and radio direction finding equipment. Also, an air-to-surface missile launching capability is planned as a future capability. Another possible addition to provide greater system flexibility is operation in the squint mode. In addition, a situation navigation and briefing display (Navigation Orientation Display) is partially developed and is available as a growth capability. Additional weapon system functions, such as defensive subsystem tie-in and fuel management, are also potential areas for expanding the capabilities of the AN/ASQ-28(V) BNS.

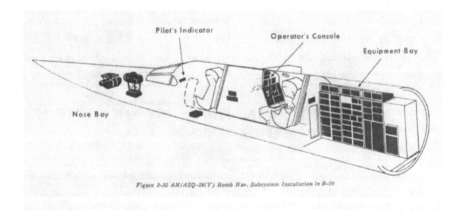

Figure 2-50 AN/ASQ-28(V) Bomb Nav. Subsystem Installation in B-70

APPLICATIONS:

Elements of the AN/ASQ-28(V) if not the entire system, are applicable to manned strategic weapon systems, for example, dromedary. Potentially, the system could be installed in the B-52 to provide an upgraded capability. The computer could be applicable to any system requiring in-flight data processing such as recon-strike and supersonic transport. Another possible application could be the Airborne Command Center.

FEATURES:

Reliability & Maintainability
Advanced fault location techniques permit isolation of malfunctions to a specific unit.

a. Automatic error checking, improved diagnostic techniques, and a random error counter minimize the effects of random and intermittent errors.

b. Voltage-mode, passive-logic computer circuits perform a maximum number of logic functions with a minimum number of transistors and basic building blocks. Silicon transistors insure reliable high-temperature operation up to 100°C.

c. During normal operation, the high-speed main computer performs all computations in parallel. If the main computer should fail, a moderate-speed emergency computer automatically
takes control and solves simplified problems serially, so that the main computer may be serviced in-flight without disturbing essential computations.

Versatility and Power
a. Circuits for converting to and from binary code and for buffering inputs and outputs are common to both main and emergency computers, so that each computer can communicate with the other and system control may be transferred from one to the other without disturbing peripheral equipments.

b. The compact magnetic drum provides storage for 26,560 instructions and 6,640 tabular data quantities for the main digital computer and 3,456 instructions and constant words for the emergency digital computer. A drum memory is used because of its nondestructive readout capability and low weight, volume, and power requirements. During normal operation the drum timing, instruction and constant tracks are non-erasable to prevent accidental destruction of the stored program and constants.

c. A 24-microsecond drum instruction time was achieved by storing instructions in a serial-parallel manner.

d. Random access core-memory selection makes possible high-speed calculations without drum look-up.

e. A high-speed, input-output processor couples the computer with peripheral equipment and calculates servo error signals to drive digital-to-analog converters.

ELEMENTS:

The DCE being developed at IBM consists of the main digital computer, the emergency digital computer, and input-output equipment.   The MDC is a high-speed, drum-core, parallel arithmetic computer. The EDC is a small, all-drum, serial arithmetic computer and is used for limited calculations in the event of malfunction of the MDC. Communication between computers is accomplished by means of an interconnection track on a common magnetic drum.   Input-output equipment provides the interconnection of the various discrete signals data, and decimal information with the other equipments of the BNS.

Controls and Displays Equipment

The CDE, which provides communication between the BNS and the operator, is being developed at IBM as an integrated console. Controls are designed by functional priority and utilization, rather than by equipment grouping, to improve operator efficiency and to provide savings in console area.

There are provisions for inserting and displaying operationally significant data and for selecting alternate modes of operation.

Radar Display Equipment

The RDE being developed at IBM consists of the electronics necessary to present radar video information to the operator in the forward-looking and high resolution sidelooking modes.   Stored presentation for both modes is available.   Display variations include plan position indicator (PPI), offset, sector scan, radar-beacon, beacon, and indirect bomb damage assessment.

Interconnection Equipment

The TCE being developed at IBM consists primarily of digital-to-analog-to-digital   converters   required   to interconnect the digital computer to the other equipments of the BNS.

Also included in this equipment group are analog computing servos, analog data transmission servos, a-c excitation equipment, and cables required to interconnect the BNS equipment.

Electronic Power Supplies

The EPSE provides the d-c power for all the BNS equipment. The power supplies for the SIE and DRE are being developed by the Autonetics Division of North American Aviation and the General Precision Laboratory respectively. The remaining three power supply units are being developed at IBM.

Radar Sighting Equipment

The RSE, being developed at General Electric, provides long range terrain sensing information under all weather conditions. The radar operates at X-band frequencies and has a choice of several operating modes. Pulse-to-pulse tunability with jam sensing and reprogramming are included along with other more conventional antijam techniques to provide a high degree of immunity to electronic counter-measures. The Range Gated Doppler Data Processor portion of the RSE is being developed by Goodyear Aircraft Corporation and extends the RSE capability to include high resolution side-looking terrain sensing.

Doppler Radar Equipment

The DRE, being developed by the General Precision Laboratory, provides velocity, elevation angle, and train angle inputs to the BNS. These inputs, when processed by the digital computer, provide true air vehicle velocity with respect to the earth for in-flight erection of the SIR and for inertial damping after erection. The Doppler radar is a high duty ratio pulsed radar with special beam lobing techniques. The DRE provides highly accurate velocity inputs over land and water without requiring landwater calibration switching.

Stellar Inertial Equipment

The SIE, designated N3B are being developed at the Autonetics Division of North American Aviation, is the

primary velocity and attitude reference for the BNS.   The
SIE consists of a high- performance inertial system with an
integral stellar tracker.   The stellar tracker, when
integrated with the digital computer, is capable of day or
night operation and provides in-flight alignment, automatic
gyro biasing, and bounded-error navigation.

Chronometer

A portable CE, developed by Itek Laboratories , operates
with the digital computer and furnished accurate Greenwich
sidereal time and precision frequency references for the
BNS.

AN/ASQ-28(v) MAIN DIGITAL COMPUTER (MDC)

The computer was developed at Owego under the B-70
Program.   A prototype was completed, evaluated, and
integrated with the other equipments of the B-70 Bombing,
Navigation, Missile Guidance Subsystem.

DFSCRIPTION:

The AN/ASQ-28(V) Main Digital Computer is a compact,
high speed, all-transistor digital computer designed and
developed as the powerful central computing element for the
bombing navigation and missile guidance subsystem of the B-
70 "Valkyrie. This general-purpose parallel computer, with
both random access and drum memories, is capable of rapid,
continuous, real-time solutions to problems associated with
automatic weapon delivery and navigation to any point on
earth with extreme accuracy, day or night, under all
weather conditions, at high speeds and high altitudes.

Reliability, maintainability, and flexibility received the
utmost consideration in every design phase.   A new
approach to fault location enhances computer maintainability
and permits in-flight repair, while flexibility is achieved by
means of unique, high-speed input-output processing
equipment.

FEATURES:

a. Replaceable (pluggable) subassemblies.

b. Rugged environmental resistance.

c. All circuits use silicon-type transistors and designed for ultra-reliable operation from 0 to 100° C

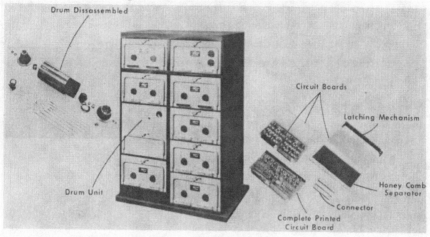

Figure 2-51 Over-all View of Main Digital Computer

d. Compact magnetic drum stores 26,624 instructions and 6656 tabular data quantities.

e. Drum also serves as the internal computer clock at frequency of 166.4 kilobits per second. Air floated drum heads provide reliable readout signals.

f. Drum tracks are noneraseable to prevent accidental destruction of stored program (drum is 6 inches in diameter and 12 inches long).

g. Unusually powerful MDC section obtained through use of a random access memory with a capacity of 1024 data words.

h. Magnetic cores and high-temperature, silicon-type transistors allow an operation cycle of 24 microseconds for a 24 bit, parallel word (memory is 9-1/2 inches long, 3-1/2 inches wide, and 5-3/4 inches deep).

i. Modular units replaceable (rack and panel connectors).

j. Intercommunication (common drum track link to EDC).

k. Error checking, through built-in circuitry and programming techniques, minimizes effects of random and intermittent errors on system performance.

l. A new hardware approach to fault isolation, supplemented by simplified diagnostic programs, permits rapid localization of computer failures.

m. Unique high-speed input-output processor with repetition rate greater than the computation cycle of the basic computer.

n. Capability of 96 programmed discrete inputs for program branching and 86 programmed discrete outputs for system control.

CHARACTERISTICS:

Type - General purpose, stored program.

Number System-Conventional binary.

Storage - Random Access Data – 1024 words Magnetic Drum.

Program Instructions - 26,624 words Magnetic Drum Tabular Data - 6656 words.

Word Length - Instructions-16 bits (including parity)

Data - 22 bits (plus sign and parity)

Arithmetic - Parallel

Memory - Magnetic Drum for Program and Tabular Storage (6000 rpm)

Random Access Data Storage

Clock Frequency - 166.4 kilobits per second

Track Density - 23. 8 tracks/inch

Logic - Four clocks per bit Number of Operations – 14.

Instruction Time - 24 microseconds (instructions stored 4 serial x 4 parallel (single-address)

Intermediate Data Access Time -24 microseconds

Operation Rates - 42,000 additions per second (24 microseconds); 3800 multiplications per second (264 microseconds)

Control Input-Output - 96 discrete inputs (contact sense) 80 discrete outputs (contact operate)

Data Input-Output - 52 variables, processed at 5200 operations per second

ELEMENTS:

One magnetic drum unit; six central computer units

consisting of program read, computer control, timing, core memory, arithmetic section, and high speed input-output processor; three input-output units consisting of discrete input and output logic, counters, converter selection and special input-output logic; diagnostic and test panel, and power module.

Hi-speed input-output processor (5200 operations per second - 52 parameters maximum) Digital servo loop may be updated at 10 millisecond intervals.

Inputs

24 shaft-to-digital 10 pulse-to-digital

1   decimal insert (manual -7 decimal digits)

96 discrete (switch or relay)

Outputs

24 digital-to-shaft 12 digital-to-pulse

2   decimal displays (7 decimal digits)

1 decimal printer (7 decimal digits)

SO discrete (relay, indicator lamp, voltage level)

CIRCUITRY:

Silicon nonsaturating voltage mode-diode logic

COMPONENTS:

1700 silicon transistors 24,000 total

Packaging (See Figure 2-52)

Modular (functional) units (4300 components per cubic foot)

Two-board unit subassemblies: 12 (maximum per unit)

Printed circuit boards (approximately 26 circuits per board)

Weight: approximately 240 lbs. (including input-output processor and drum) Drum = 48 lbs.

Volume: 7.4 cubic feet (same as above)

Operating Environment: Supersonic military aircraft

Ambient temperature range: 0-100° C (forced air cooled)

ADVANCED CAPABILITY RADAR

The ACR System was produced for use in the B-52 Weapons System.

DESCRIPTION:

The ACR terrain clearance system is a radar information processing and display system development by the IBM Space Guidance Center for the B-52 aircraft. Utilizing the accuracy and data rate inherent in monopulse radar, this system equips the B-

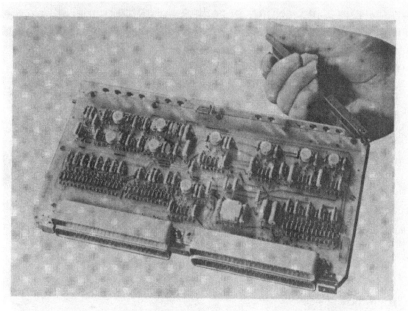

Figure 2–52 Printed Circuit Sub-Assem-AN/ASQ–28(U)

52 with a low level flying capability by providing the pilot with an obstacle warning system.

The ACR system is a multi-mode, flexible system constructed from basic "building-blocks" which can be easily re-arranged and adapted for use with weapons systems other than the B-52. The present system capabilities are classified.

APPLICATIONS:

In its present or revised forms, the ACR system is applicable to low level flight and obstacle warning provisions for the following applications:

Strategic bombers   Tactical   bombers   Fighter   aircraft
Transport aircraft

Ground support aircraft

FEATURES:

Simultaneous navigator (radar ground map) and pilot (terrain clearance) displays.
Selectable terrain clearance displays:
    ** plan (range vs. azimuth)
    ** profilometer (elevation vs. azimuth)
Selectable stabilization modes:
    **horizontal (pitch stabilized and roll compensated)
    ** aircraft flight vector (direction aircraft is moving)
    ** aircraft fuselage (direction aircraft is pointed)
Selectable flying levels (high data rate):
    ** computation at radar speeds
    ** one second display updating
Built-in failure warning circuitry

ELEMENTS:

The packaging and instrumentation of the ACR system has been modularized according to basic functions to simplify maintenance and to form basic building blocks for alternate systems. Built-in test and alignment features have also been utilized throughout.

The ACR system consists of three basic subsystems.
    ** Monopulse radar
    ** Terrain obstacle computer
    ** Information display equipment

TOPOGRAPHICAL COMPARITOR

Facility: Owego
Status: 500-700 Delivered to Air Force
Applications: B-52 Aircraft

DESCRIPTION:

The Topographical Comparitor is a map-matching device which enables an aircraft operator to compare the present position information of the forward-looking radar with the store reference information. The comparison provides an indication of the error in the position of the aircraft.

The reference information is stored on 35 -mm film.   It may be radar screen photographs from previous flights, photographs produced by a radar simulator, or imitation radar pictures drawn by artists using topographical maps as guides.   These reference photographs may be modified optically to permit use at an altitude different from the original data acquisition.

The error correction is performed manually and is proportional to the signal given the computer representing flight path error at the optimum map-match.   The optimum map match may be obtained with or without flicker.   The system is equally adaptable to direct or offset target techniques.

NAVIGATIONAL DATA RECORDER

Produced at Owego Space Guidance Center.

DESCRIPTION:

The Navigational Data Recorder is a compact airborne numeric printer developed for the B-70 Bombing Navigational System AN/ASQ-28.   The system requirement is to present a progressive readout of computed information that could be viewed by the navigator on a 1-1/4 inch, vertically fed paper tape (See Figure 2-70.)

APPLICATIONS: B-70

The existing printer can be modified for any commercial or military application requiring up to 16 characters by changing the letters and numerals on the print cylinder. The basic design principles incorporated can be extended to the design or recorders for applications requiring greater than 16 characters print-out.

FEATURES:

This printer was designed to provide a continuous display of present position latitude and longitude data under control of the computer.   The single-element print cylinder can be altered to read out any information that can be interpreted from a 4 channel, binary input.

CHARACTERISTICS:

The recorder was designed to meet Mil Spec E-5400C modified by crew bay conditions in the B-70 air vehicle.

Figure 2-70 Navigational Data Recorder

Print time - 0.5 sec per character line (1200 characters per minute).

Horizontal line length - 8 characters at 0.1 inch spacing.

Character height - 3/32 inch (also modeled at 1/8 inch).

Characters - 10 numeric, fi special Vertical spacing - 6 lines per inch.

Viewing field - 10 lines

Signal input-Binary Coded Decimal to four set-up magnets (transistor drive).

Cycle - common pulse from computer.

Set-up and cycle clutch magnets -28 VDC.

SAGE (SEMIAUTOMATIC GROUND ENVIRONMENT) SYSTEM

Out of production. Twenty-three systems fabricated and installed

DESCRIPTION:

The SAGE system is the major surveillance and control system in the total United States Air Defense Complex. Its nerve centers are the Direction and com -bat Centers which house the SAGE computers. These computers organize large quantities of diverse information from a large number and variety of sources. This data includes short-range and long-range radar returns from land and offshore (Texas Towers) installations and from Picket ships and planes. The computers also receive weather reports, aircraft flight plans, and weapons status reports. (See Figure 2-1)

CAPABILITIES:

The computers process and display the above data for Direction Center personnel to track and identify aircraft. The computers also assist the operators in selecting and directing weapons, and keep all personnel fully informed on the air situation. When certain actions are required, the computers will alert personnel, will display possible actions to take, and, after action has been taken, will continually compute and display all subsequent changes to the original data.

APPLICATIONS:

Air Defense
Air Traffic Control

The Direction Centers

The Direction Center is a data processing and weapons control installation responsible for the Air Defense of a geographical sector. Each Direction Center houses a SAGE (AN/FSQ-7) Computer System which is composed of two digital computers and input/output and display equipment. This installation, known as a duplexed SAGE Computer System, insures reliable around-the-clock Air Defense. Communication facilities are provided between Direction Centers, and between Direction Centers and Combat Centers. A Direction Center also includes power and air conditioned buildings, fuel storage, and supporting facilities.

The Combat Centers

Combat Centers have been provided to monitor and supervise groups of Direction Centers. To the Combat Centers come Air Defense data from the subordinate Direction Centers. From this information, suitable displays of the current air defense situation are presented, enabling the Combat Center Commander to exercise over-all supervision and control of Air Defense in his division. The Combat Centers are equipped with an AN/FSQ-8 Computer System which is similar to the AN/FSQ-7. The Combat Center Computers have less input/output equipment and displays because they handle less data than Direction Centers, The Combat Center is located jointly with one of its subordinate Direction Centers.

ELEMENTS:

The Central Computer

The SAGE Computer is a stored-program, general-purpose, digital computer with special features designed to cope with the problems of Air Defense. The dual arithmetic unit permits simultaneous (parallel) operation of two numbers. This feature increases the effective speed of the computer and permits the calculation of the direction-distance vectors of targets.

The greater use of multiplication in Air Defense problem-solving has been provided for by a multiplication speed of 16. 5 microseconds, much faster than is common for such computers. Calculations in real time (concurrent with the air situation) have been made possible by interruption facilities which allow the Computer to continue to operate during Input/output operations.

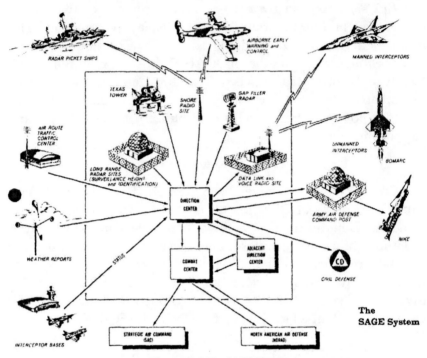

Figure 2-1 The SAGE System

The Central Computer is the designation commonly given to the units of the SAGE Computer concerned with instruction processing and computation. The SAGE Computer is composed of units, modules, and pluggable units. Each Computer is divided into approximately 30 units (excluding consoles), each unit made up of an average of 10 modules. The modules, in turn, are composed of approximately 20 pluggable units that can be quickly replaced. Spare pluggable units have been provided for future expansion.

Each module acts as a separate duct for conditioned air that comes into the bottom of the modules. The air is exhausted through openings surrounding the tubes, thereby cooling them.　Each opening is designed according to the cooling requirements of the particular tube used.　Power can be removed from an individual module for maintenance.

Magnetic Core Memory Array

The magnetic core memory is the high-speed storage device for the SAGE Computer.　Information can be read into or out of this memory in six millionths of a second.

The memory array is the unit in which the information is stored.　Binary information (using only 1's and 0's) is stored in minute ring-shaped iron cores by magnetizing each core in one of two possible directions.　The memory array is composed of 33 horizontal planes stacked vertically, each plane containing 4,095 magnetic cores.

An important advantage of the magnetic core memory is its random access; i.e., information can be stored and retrieved in any sequence,　Compared to other types of storage the core memory is compact and requires no periodic regeneration of stored information.

Increased Capacity Memory

SAGE capability has been multiplied many times by the addition of an Increased capacity core memory.　The new memory is 16 times the size of the original memory, with storage capacity for approximately 65,000 words of information.　The new memory has accelerated the Air Defense Program by providing storage capacity for many more of the commonly used routines and tables.　Also, much larger blocks of information can be transferred at a time from the drums to memory, saving much of the time that had been formerly consumed in address searching and information transfer.

Duplex Operating Console

Each SAGE Computer System has an operating console for each of the two computers, with controls and indicators for starting, loading, monitoring, testing, and stopping the Computer.　Hundreds of neons indicate the condition of Computer circuits, including the contents of arithmetical circuits.   Switches and buttons on the console can be used to direct the Computer to automatically process Air Defense information or to execute a maintenance or marginal checking program.

The marginal checking program is a unique preventive maintenance routine devised to meet the extreme reliability requirements of Air Defense.   While one computer is processing Air Defense data, maintenance operations are being carried out on the Standby Computer, Circuits are artificially aged by the varying of supply voltages, making it possible to locate marginal, components and replace them before they fail.

Simplex Operating Console

The simplex operating console is vised to monitor and control communication lines into Direction Centers.   It is made up of 72 pluggable control panels and 3 fixed panels, each with appropriate switches, indicators, and alarms. Channel control panels (from gap-filler and long-range radars, and crosstell lines) compose the greater part of the console; in addition, there are control panels for power, marginal checking, the computer entry punch, oscilloscope and probe equipment, and intercommunication equipment.

Duplex Switching Console

The Duplex switching console (one for each Computer) contains the controls required to switch Air Defense operations from one Computer to the other.   Each duplex switching console is provided with indicators and alarms for showing the status of equipment.

Input/Output Equipment

An outstanding characteristic of the- SAGE Computer is the input/output equipment that enables the Computer to receive and transmit information through the wide variety of communication equipment of the SAGE Air Defense System.

The   demands   of   around-the-clock   Air   Defense presuppose   extensive   tests,   diagnosis,   and   preventive maintenance of the Standby Computer. This is carried out largely with card and tape equipment, although high-speed drums and displays are generally reserved for operational use.

IBM punched cards are used to enter data into SAGE Computers because of the flexibility of card entry and the availability of associated keypunching, verifying, and duplicating equipment. Important information is verified prior to computer entry.

Program entry can be expedited by the simultaneous preparation of data on several keypunches, and the cards provide a permanent record of the data.

Computer Entry Punch - Type 020. The Computer entry punch is the manual input unit to the SAGE Computer. Information such as weather reports or flight plans can be entered either directly from a keyboard or from previously punched cards. Besides punching cards and reading them into the Computers, the unit can also print the punched information across the top of the cards for ease of reading.

The combination alphabetic and numeric keyboard has 44 keys, 11 of which are for special characters.

CARD Reader - Type 713. The card reader fs used to enter instructions and data into the SAGE Computer. The card reader is a semiautomatic machine that is controlled by the Computer. Information is read into the Computer at the rate of 150 cards per minute.

Magnetic Tape Drive - Type 728. The SAGE Computer is equipped with eight tape drive units, each capable of storing or reading out over a million words of information on each reel of tape at a rate of 520 words per second. Data is recorded in the form of small magnetized areas on the tape and may be stored in this form indefinitely. The tape may be erased and re-used as desired, making it an economical storage medium. Except for the manual loading and removing of the tape reels, the tapes are fully automatic once they are placed under the control of the Computer.

QF1 **Mapper Console.** Each mapper console receives all the radar returns from the gap-filler input site. This information contains returns from objects other than targets, such as mountains, buildings, ground clutter, cloud banks, stray noise, deliberate jamming, etc. The irrelevant information is eliminated by mapping it out with an opaquing substance so that only the desired target information is picked up from the mapper screen by the over charging photoelectric pick-up unit and allowed to enter the Computer.

Magnetic Drums. The SAGE Computer Has 12 main drums and 12 auxiliary drums. These are used as auxiliary memories for the Computer, The access time of data stored on drums, although slower than the access time of data stored in Computer memory, is appreciably faster than the access time of data stored on card and tape machines. In addition, the drums serve as time buffering devices. Information passing between the high-speed Computer and slower-speed input/output devices is stored temporarily on the drums. This permits the optimum use of time by both the Computer and the slower devices.

Card Punch - Type 733. The card punch is an automatic device designed to punch standard IBM cards with up to 24 Computer words, at a rate of 100 cards per minute. The card punch is used to maintain a permanent record of the operational and program results of the SAGE Computer.

Line Printer - Type 718. The line printer is a device for recording information from the SAGE Computer in printer form. It has 120 alphabetic-numeric type wheels, each consisting of the complete alphabet, numbers, and nine special characters. The printer operates at 150 lines per minute.

Tape-to-Card Printing Punch - Type 017. Some of the Air Defense information arriving at the Direction Center is in the form of Teletype punched paper tape. The tape-to-card punch is used to transfer this data to IBM cards, which are then read into the Computer. In addition to punching cards, the punch can also print the punched information across the top of the cards for ease of reading.

Teletypewriter Monitor. The teletypewriter monitor is used to test the teletypewriter storage section of the SAGE Computer. This section temporarily stores data from the

Computer before it is transmitted over 25 channels to tele-typewriters located at antiaircraft control centers, air bases, higher headquarters, etc.   Messages are transmitted at a nominal rate of 50 words per minute.

The SAGE Computer accepts Air Defense intelligence from radar outposts and other data collection points.   After being processed by the Computer, this information is presented on visual displays where it is interpreted by Air Defense personnel. Although there are two basic consoles, situation display and auxiliary, there are variations in their controls and in the information which they display.   The information shown on each console is determined by the console's A i r Defense function, whether for air surveillance, aircraft identification, or weapons' direction.   Indicators, controls, and communication equipment permit operator response to an Air Defense situation.

Situation Display.   The 19-inch situation display shows an ever changing plan-position graphic display of the changing air situation with correct geographical relations between fixed points and moving targets.   Additional descriptive data in the form of vectors and characters (letters, numerals, and special symbols) is posted adjacent to particular points and targets. A message pattern depicts four basic types of messages that may be displayed: radar data, track data, track data tabular information, and track data vector messages.

The 5-inch digital display primarily furnishes additional information to supplement situation displays and to summarize existing situations, such as weather and assignment data.   The data appears as a tabular array of characteristics (letters, numerals, and special symbols, but no vectors) only on the viewing screen of the cathode-ray tube.   These characteristics appear in 2 columns of rows with 5 characteristics to the row and 1(> rows in a column.   A digital display, unlike a situation display, is changed only when the Computer orders a new display or when additional information is requested by the operator.

Situation Display Console.   The situation display console is   used   for   air   surveillance,   aircraft   identification, monitoring, and weapons direction where a plan-position type of display is needed.

It is equipped with a situation display, various indicators and controls, a light gun, and usually a digital display. It may also have one or two keyboard control panels, commonly called side wings.

The display console is equipped with expansion, off-centering, brightness and communication controls. In addition, display controls enable operating personnel to select or reject certain types and categories of information according to the tactical requirements assigned to each operating position.

The side wings have pushbuttons for setting up messages, which are then entered into the Computer by a button or the light gun. The light gun, a hand-held photo-electric device, identifies a target on the display screen to the Computer,

Auxiliary Display Console. The auxiliary display console is used for displaying certain categories of information where a tabular rather than a plan-position type of display is needed. It is equipped with alarms, warning lights, intervention switches for inserting information into the Computer, and usually one digital display. Some of the auxiliary consoles, however, have no display, but are used only to insert information into the Computer. An auxiliary console may be used in conjunction with a situation display console, permitting a 2-man team to carry out an Air Defense task.

File Initiation Area Discriminator. Mounted on Situation Display Console, the file initiation area discriminator is, in effect, a giant light gun which can monitor an entire situation display. The initiation area discriminator is used in the Air Surveillance Room to automatically cause the Computer to create tracks from radar data. Opaque material is applied to the face of the display to block out the areas where automatic initiation is not desired. Whereas the light gun is hand-held the area discriminator is rigidly mounted at a fixed distance from the situation display so that its viewing field can encompass the entire display area.

Situation Display Camera Console. The Situation Display camera console is used to make a permanent 35-mm photographic record of situation displays. These re-records are used for tactical analysis and training aids. Two camera consoles are used at a site: one for active displays, the other for standby displays. The semiautomatic camera is controlled by the Computer.

Photographic Recorder - Reproducer. The photographic recorder-reproducer automatically photographs the air situation presented on a special situation display, develops the film, and projects an image on a screen at the Command Post. A new display is recorded and reproduced every 30 seconds, providing command level personnel with an over-all view of the air situation as it progresses.

AN/FSQ-7 COMBAT DIRECTION CENTRAL

23 systems fabricated and installed. Out of production.

DESCRIPTION:

In October 1952, IBM was awarded a contract to assist the Lincoln Laboratory of the Massachusetts Institute of Technology in the design and construction of the AN/FSQ-7 Combat Direction Central. Later, the U.S. Air Force awarded a contract to IBM for the construction and installation of AN/FSQ-7 Computer at strategic points within the continental United States as part of the Semiautomatic Ground Environment (SAGE) System. The Junction of the AN/FSQ-7 Combat Direction Central is to process data relating to the detection, identification, and tracking of aircraft and to provide information for the direction and control of air defense weapons. (See Figure 2-9)

APPLICATIONS:

Data processing and weapons control computer for the SAGE System.

FEATURES:

The AN/FSQ-7 Computer *is* a stored-program, general -

purpose digital computer with special features designed to cope with the problems of air defense. The dual arithmetic unit permits simultaneous (parallel) operation on two numbers. This feature increases the effective speed of the computer and permits the calculation of the direction-distance vectors of targets. The greater USE of multiplication in air defense problem-solving has been made possible by interruption facilities which allow the computer to continue to operate during I/O operations.

CHARACTERISTICS:

Central Computer
 Word size 33 bits
 Memory: 69,632 words
 Basic timing frequency: 2 mc
 Memory cycle: 6 usec
 Number of basic instructions: 48
 Average instruction rate: 85,000/sec
 Arithmetic subtract time: 12 usec
 Transfer rate (consecutive rate); 10 usec/word
 Arithmetic multiply time: 16.5 usec
 Arithmetic divide time: 51usec
 Two Magnetic Core Memories: 64^2 and 256^2
 64^2 Memory; 4,096 words – 33 bits/word including 1 Party Bit
(See Figure 2-10)
 Vacuum tube circuitry
 Weight, 7,718 lbs.
Drum System
 Type of recording: Discrete spot
 Number of drums: 12
 Number of fields/drum: 6
 Total number of registers: 153,000
 Maximum access time: 20 ms
 Drum Diameter: 10.7 inches
 Weight: 105 lbs.
 Length: 12.5 inches
 RPM: 2914
Input System
 LRI channels: 36 maximum (32 plus 4 spares)

XTL channels: 12 maximum (11 plus 1 spare)

GFI channels: 18 maximum (16 plus 2 spares)

Output System

Outputs: Ground-to-air

Ground-to-ground

Teletypewriter

Display System

Type of display: PPI (SD)

Tabular data (DD)

Number of consoles: 110 (SD) 124 (DD)

Magnetic Tape Storage:

Number/computer: 4

Read time: 75 inches/sec

Write tine: 75 inches/sec

Maximum word storage: 1,200,000/reel

Time to read or write full reel: 6-1/2 minutes

Local Peripheral Equipment;:

Card reader: 150 card/min

Line printer: 150 lines/min

High-speed punch: 100 cards/min

Display cycle: 2.56 seconds (SD) - 1.34 seconds (DD)

Marginal Check Voltages: +250, +150, +90, -150, -300V d-c

Power System:

Power available: 3000 kw

Voltages: Unregulated a-c: 3-phase, 60 cycle, 120/208V

Standard d-c: +600, +250, +150, +90, +72, +10, -15, -30, -48, -150, -300

Nonstandard d-c: +270, +140, -60, -130, -270

Cooling Requirements: Forced air at 25° C

Space Requirements: Three-floor building; each floor 270 ft. x 150 ft.

Total Weight: 250 tons

The AN/FSQ-7 Computer is composed of units, modules, and pluggable units. Each computer is divided into approximately 30units (excluding consoles); each unit comprises an average of 10 modules. The modules, in turn, are composed of approximately 20 pluggable units that can be quickly replaced.

The main subsystems of the AN/FSQ-7 Computer:

Central Computer System

Drum System
Display Input System
Output System
Power System

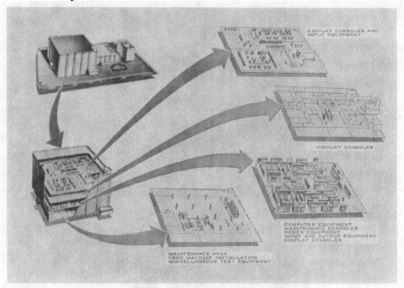

Figure 2-9 AN/FSQ-7 Combat Direction Control

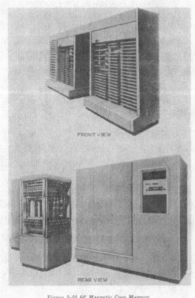

Figure 3-10 64' Magnetic Core Memory

Figure 3-11 256' Magnetic Core Memory

REFERENCE DATA:

Theory of Operation manual, AN/FSQ-7
Combat Direction Central and AN/FSQ-8 Combat Control Central

AN/FSQ-31(V) DATA PROCESSING CENTRAL

Produced under AF contract. Out of production.

DESCRIPTION:

The AN/FSQ-31(V) is a solid-state data processing central (DPC) produced for SAC. This computer operates on 48-bit words and can process as many as 400,000 instructions (single-address) per second. (See System Layout, Figure 2-1 a)

APPLICATIONS:

Part of a global system for the planning, direction and control of SAC operations 465-L.

FEATURES:

The AN/FSQ-31(V) is a large-capacity data processor that features reliability, versatility, and expandability which is largely obtained by adding modules. It has solid-state circuits that are packaged in Q-Pac's, the basic machine building blocks. The number of Q-Pac types has been held to a minimum to facilitate maintenance and logistics.

Programming Features — Powerful programming tools have been included in the machine design to increase data-handling efficiency and ease the programmer's task. Instruction-modifying bits permit the use of full-or-half-word logic and provide byte control; i.e., the selection and positioning of items of information which do not require a full word. The programmer's task is simplified by three levels of addressing: Direct Addressing (ADR), Immediate Addressing (R), and Indirect Addressing (I); and by three methods of indexing: Direct Indexing (IX), Double Indexing (DI) (IX), and Relative Indexing (IX).

Central Processing Unit —This unit features an interrupt

control element whereby (a) single-bit errors are automatically detected and compensated for, so that the operational program can continue and (b) priority can be given to the more important incoming messages.

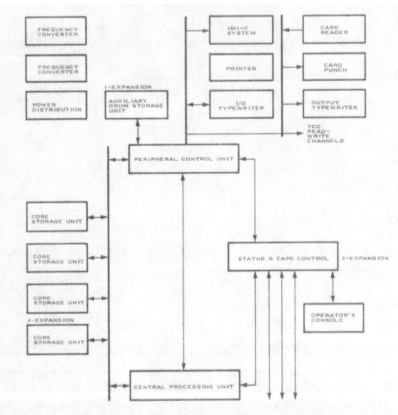

Figure 2-13 AN/FSQ-31(v) Data Processing Central (System Layout)

Core Storage Units — There are four core storage units (expandable to 8), each of which provides storage for 16,384 computer words.   These units are independent and capable of simultaneous operation with a compatible computer such as the AN/FSQ-31.   In the AN/FSQ-31(V), all four units can operate simultaneously, two with the CPU and two with the PCU.   The combination of simultaneous memory operation, 2-level instruction decoding, and independent I/O transfers (see paragraph below) makes it possible to do the following in one memory cycle: (a) overlap the processing of two instructions and (b) execute transfers between core storage and one high-speed and one low-speed I/O device. This is a major feature.

Peripheral Control Unit — This unit regulates all I/O operations BO that one high-speed device (e.g., drums) and one slow-speed device (e.g., tapes) may simultaneously exchange data with core storage independently of the CPU. The slow-speed devices time-share control circuits to provide continuous operation of all tape channels (1-3, expandable to 6), printer or punch, card reader, I/O typewriter, output typewriter, and two Traffic Control Center channels.

Auxiliary Drum Storage Unit — Each DPC has one such unit, with provisions for expansion to two. Each unit has two drums (expandable to 4) and the associated controls to provide storage to 34 fields of 8192 words. Thus, there is capability for storing over one million words at an approximate transfer rate of 2.75 usec per word.

Status and Tape Control Unit — This unit contains circuits required for the operator's console and the circuits of the tape adapter element. Each tape channel may operate with one to eight tape drives to transport information at either of two densities (200 or 555 characters per inch).

CHARACTERISTICS:

Numerical System: Binary
Core Storage Element (computer memory)
    Cycle time: 2.3 usec
    Words per unit; 16,384
Number of units: 4-8
Number of Instructions: 128 possible; 69 assigned with various modifiers to effectively expand this instruction capability.
Clock Rate: 6.41 mc
Machine Cycle: 2.5 usec
Time Pulses: 16 at 156-musec intervals
Instruction Rate: Peak - 400,000/sec
Instruction Word Length: 48 bits plus 2 parity bits
Auxiliary Drum Storage Units:
    Number of units: 1 -2
    Drums per unit; 2-4

Storage per drum: 17 fields of 8192 words (139,264 words)
Word transfer rate: 2.75 usec (approx.)
Local-Input-Output
Tape units: up to 48
Tape speed: 62,500 characters/sec
Card punch speed: 100 cards/min
Card reader speed: 250 cards/min
Line printer speed: 600 lines/min

ELEMENTS:

A central processing unit (CPU); four core storage units, expandable to eight; a peripheral control unit (PCU) which controls all I/O operations; one auxiliary drum storage unit, expandable to two; a status and tape control unit; card reader; printer; card punch; I/O typewriter; and Traffic Control Center read-write channels.

REFERENCE DATA:

Performance specifications:

SC-10-1	SC-15-1	SC-34-1	SC-52-1
SC-11-2	SC-16-1	SC-35-X	SC-63
SC-13-1	SC-17-2	SC-51-1	SC-64
SC-14-1	SC-19-2		

AN/FSQ-32 COMBAT CONTROL CENTRAL

One engineering prototype fabricated under AF Contract AF-30<635)-12756. Delivered to SDC Santa Monica, California. Due to contract cancellation, two additional systems were converted to Q-31 configuration and delivered to the SACCS program. Out of production.

DESCRIPTION:

The AN/FSQ-32 is an air defense control central designed to meet combined air-defense, air-traffic control needs. It is compatible with the SAC AN'FSQ-31(v) Data Processing Central. (See Figure 2-14)

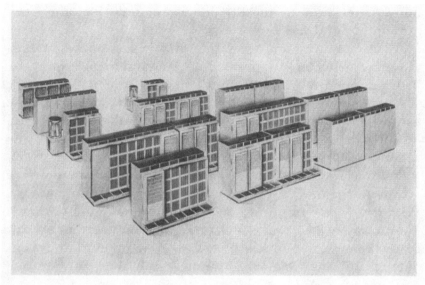

Figure 2-14 AN/FSQ-32 (XD-1) Artists Rendering

APPLICATIONS:

Air Defense System.   With additional refinement, system can operate as a combined air defense, air traffic control center.

FEATURES:

The AN/FSQ-32 is a large-capacity computer system that features reliability, versatility, and expandability which is largely obtained by adding modules.   It has solid-state circuits that are packaged in Q-Pac's, the basic machine building blocks.   The number of Q-Pac types has been held to a minimum to facilitate maintenance and logistics.

The programming features and the features of the central processing unit, the peripheral control unit, and the core storage units are essentially the same as those of the AN/FSQ-31.   In the AN/FSQ-32, however, the interrupt control element contains detail condition registers whose bits continually indicate the operational conditions of various machine elements to enable the program to adapt to varying situations within the machine.

Auxiliary Drum Storage Control Unit — The AN/FSQ-32 has one such unit that controls the operation of one magnetic drum, but has the capacity for controlling as many as four. For further details, see the Transistorized Magnetic Drum System.

Real-Time Input Unit —This unit is a special feature of the AN/FSQ-32. It performs bit-by-bit processing that utilizes an input core memory for assembly and storage. The core storage allocation for each input channel is variable, assigned by control words associated with the channel. The input data is available to the central processor either by directly addressing the input core storage or by block-transfer operation. A new feature is a low data-rate section which automatically processes incoming teletype messages. With this input unit, the AN/FSQ-32 can have as many as 76 input channels, each of which is capable of receiving its data from either of two phone lines (active or backup) associated with it. The phone lines use Bell A.I Data Service. In addition, the AN/FSQ-32 can accept information from a maximum of 64 teletype lines, each of which may be operated with a 5-bit code at 60, 75, or 100 words per minute or with a 6-bit code at 53, 66 or 88 words per minute.

Real-Time Output Element —All output messages can now be stored on a single output-buffer field of the DATOR drum. Ground-to-ground and ground-to-air information can be sent out on as many as 33 channels, each of which can supply serial data to either or both of two phone lines associated with it. The phone lines use Bell A.I Data Service. In addition, the AN/FSQ-32 can supply output data to as many as 50 teletype channels that operate at 60 words per minute with a 5-bit code. With some design effort, the system could be expanded to accommodate a 6-bit code.

Advanced Display System — This system features consoles designed to facilitate man-machine communications. The incorporation of the "bright tube" permits better man-machine communication because a higher level of ambient lighting is permissible. Both tabular and symbolic data may be viewed upon the bright tube.

CHARACTERISTICS:

Numerical System: Binary, parallel operation

Core Storage Element (computer memory) 128^2
  Cycle time: 2.3 usec
  Words per unit: 16,384
Number of units: 4-8
Machine Cycle: 2.5 usec
Instruction Rate: Up to400,000/sec
Instruction Word Length: 48 bits plus 2 parity bits
Arithmetic Speeds (fixed-point)
  Add or subtract: 5.0 usec; 2.5 usec if overlapped
  Multiply: 9.6 - 31 usec
  Divide: 40 usec
Arithmetic Speeds (floating-point)
  Add or subtract: 5 usec; 2.5 usec if overlapped
  Multiply: 14 - 58 uaec
  Divide: 70 usec
Core Storage Element (input memory)
  Cycle time: 2.3 usec
  Words per unit: 16,384
  Number of units: 1 - 2
  Size: Height: 6'8"; Depth: 3'1"; Width 9'1"
  Weight 6,300 lbs.
Drum Storage
  DATOR drum: 1
  Auxiliary storage drum: 1 (expandable to 4)
  Storage per drum: 17 fields of 8192 words (139,264 words)
  Word transfer rate: 2.75 usec
  Average access time: 11 ms
Peripheral Equipment
  Tape units: maximum of 48
  Tape speed: 62,500/22,500 characters/sec
  Card punch speed: 100 cards/min
  Card reader speed: 250 cards/min
  Line printer speed: 600 lines/min

In addition to the above memory systems one experimental memory "Compact Magnetic Core Memory" was fabricated for the Advanced Display Console on the AN/FSQ-32 (XD-1).

Compactness — The principal component is the core array

which is constructed of vertically stacked plane frames. As many as three of these frames can be housed in one SMS swing gate; another SMS swing gate houses associated circuitry. Thus, the complete unit consists of a double SMS gate whose outer dimensions are 13.75 inches x 15.5 inches x 22.5 inches.

Capacity -
    4096 words of 24 bits
    2048 words of 36 bits
    1024 words of 54 bits
Memory Cycle - 6 usec
Access Time for Readout - 3 usec
IBM Type 17 Ferrite Cores – ID 0.030"; OD 0.050"
Circuits – All transistorized
Air Cooling - 25 CFM at 70 ± 1°F 140 watts heat – load
Size – 13.75" x 15.5" x 2.5"
Weight: 110 lbs.

Physically, this Magnetic Core Memory consists of one unit which supports two standard modular system (SMS) swing gates. Included in its logic are the necessary drivers and sense amplifiers, the core array, and some timing circuits. The memory is computer-controlled in the sense that the memory address register, line decoders, time pulse distributor, memory buffer, register, and power supplies are external to the memory unit.

ELEMENTS:

Central processing unit; four core storage units, expandable to eight; one auxiliary drum storage unit, expandable to two; local peripheral units such as magnetic tapes, the I/O typewriter, an output typewriter, and the 1401 processing unit, all of which are controlled by the peripheral control unit (PCU); a real-time input unit with one core storage unit, expandable to two; a DATA drum unit that serves as a buffer for both a real-time output unit and an advanced display system which has bright-tube consoles.

TARGET ACQUISITION SYSTEM (TAS)

Design of the TAS was accomplished by tin-Space Systems Dept,, Bethesda; procurement, fabrication and detailed design were provided by Kingston.

DESCRIPTION:

The TAS was designed for the U.S. Naval Ordnance Test Station (USNOTS), China Lake, California, under Contract No. N-123 (60530) 268-70A. The system was based on specifications released by USNOTS Personnel.

The Target Acquisition System (TAS) is a data acquisition and processing system designed to provide pointing instructions to operators of eight manually positioned cinetheodolites. These instructions enable expeditious viewing of airborne targets for subsequent recording on film. This system is fully automatic within the limitations imposed by the data sources. Although designed for a cinetheodolite acquisition system, the TAS combines features flexible enough for an integrated range control system, such as drone vectoring, aircraft guidance, and fire control displays.

ELEMENTS:

The Target Acquisition System is composed of four subsystems; MIDAS Input, Radar Input, Computing, and Output. The Interrelation of these subsystems is shown in Figure 2-3. The components of each subsystem are described briefly in subsequent, paragraphs.

MIDAS Input Subsystem

This subsystem presents targets-position information to the Computing Subsystem and is fully automatic to the extent that the MTDAS device is automatic. Also, this subsystem is capable ot handling all information presented by MIDAS.

MIDAS is an interformeter type of trajectory measuring device composed of two remote stations (East, and West), which alternately supply, via microwave, target(s)-position information to two microwave receivers.

Radar Input Subsystem

This subsystem provides range, azimuth, elevation, identity, range time, and on-track information to the Computing Subsystem. The instantaneous range data and antenna positional data of the tracking radars (M33, 7298, and MPS-2U) are sent to synchro transmitters and digital encoders. The digital encoders are sequentially strobed, and the encoded data (Gray code) sent, in parallel, to the 1901 Target Data Buffer.

Computing Subsystem

The Computing Subsystem receives the data from the MIDAS and Radar Input Subsystems and performs the data handling and computations necessary for the derivation of the required acquisition messages. These messages are then transmitted to the Output Subsystem for subsequent transmission to the remote stations.

Output Subsystem

The acquisition messages derived by the Computing Subsystem are addressed and then transmitted via microwave to a UHF broadcast transmitter. The acquisition messages contained in this broadcast are also received at the eight instrumented cinetheodolite stations, At these stations, the addresses of the messages are checked, and the appropriate azimuth and elevation data gated into comparators. As a result of this comparison, pointing instructions are displayed within the viewing telescopes allowing expeditious acquisition of airborne targets.

FEATURES:

The equipment available at USNOTS prior to installation of the TAS included:

** IBM 7090 without a real-time I/O device.
** Cinetheodolites
** Microwave and UHF Links
** Midas Devices and Radars

For TAS, IBM supplied the following equipments utilizing the above existing equipment.

IBM 7281 MOD II DCC

This unit operates as an integral part of the IBM 7090

system and contains sequencer, multiplexer, and control circuitry common to all subchannels of the 7281 II DCC The operation of this unit is such that the subchannels may operate independently and may employ data of different rates and formats in asynchronous modes of operation.

4901 Target Data Buffer

The circuits contained in this unit prepare the Radar

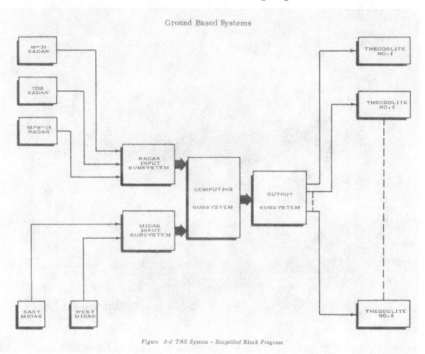

Figure 2-4 TAS System - Simplified Block Program

Input Subsystem data and the MIDAS Input Subsystem data for transmittal via microwave to the 7090 computer. (See Figure 2-4)

4902 Position Control Unit

This unit is part of the Output Subsystem, and one of these units is located at each cinetheodolite station to provide pointing instructions to the operators of tie cinetheodolites. The circuits contained in this unit accept the data (acquisition messages) from the Computing Subsystem and convert this data into pointing instructions which are displayed within the cinetheodolites. (See Figure 2-5)

Figure 2-4 IBM 4901 Target Data Buffer

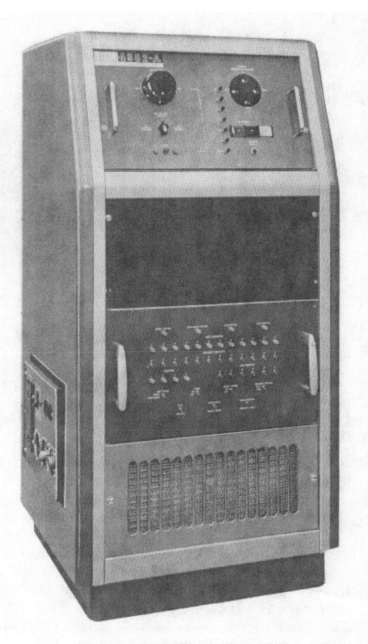

Figure 2–5 IBM 4902 Position Control Unit

(NOT SHOWN IS THE I O TYPEWRITER, CARD READER, FERRITE CORE MEMORY, AND DRUM SYSTEM)

Figure 2-12 RTA Computer

RELIABILITY TEST ASSEMBLY

Built under contract No. AF30 (635)-1404, completed 19S9. Delivered to Mitre Corp., out of production.

DESCRIPTION:

The Reliability Test Assembly (RTA) is a solid-state, less-than-full-scale computer designed to develop and test in a real computer environment concepts intended for incorporation into an advanced solid-state digital computer. The unit is a single-address, parallel system which uses germanium transistors and operates in a speed range of 5 to 10 me. Design criteria for the transistorized production computer are based on military requirements for SAGE, SAC, and other advanced computer situations. Emphasis is on reliability, speed, operational capacity, ease of maintenance, and miniaturization. (See Figure 2-12)

FEATURES:

The RTA features major improvements over other computer systems as follows: all-transistor circuitry;

operates with multiple memories, any two of which are simultaneously accessible; simultaneous memory operation with two-level decoding speed instruction rate by making possible overlapped processing of two instructions; extensive self-error-checking circuitry insures almost 100 percent single error detection capability; and space, power, and equipment requirements reduced throughout.

Circuits — eight circuits are packaged into 16 different potted subassemblies called Q-Pac's. The circuits will operate under extreme degradation conditions. For example, any circuit will still function when all its components are in worst direction to end-of-life tolerances, when input signals and power supplies are varied to worst end-of-life tolerances, and when heaviest loading conditions are applied.

Packaging — the Q-Pac is virtually immune to shock and vibration. As many as 100 Q-Pac's can be mounted in a pluggable drawer, with 16 drawers to a module. The RTA uses a maximum of 96 drawers.

Transistors — essentially, three transistor types are used throughout. A graded base germanium transistor with characteristics similar to the 2N501 is the basic component in more than 85 percent of the machine.

Error Detection — error-checking logic is sufficient to insure 100 percent detection of all single errors.

I/O Operation — the simultaneous operation of a variety of devices is possible: high-speed magnetic drums, magnetic tapes, and I/O typewriters. A peripheral unit processor reduces standard mismatch between internal computing speed and external I/O transmission rates.

CHARACTERISTICS:

Numerical System: Binary, parallel operation
Number of Instructions: 32 possible; 23 assigned
Clock Rate 6.25 mc
Memory Cycle 2.0 usec
Machine Cycle 2.24 usec (RTA is also compatible with a 1-usec, 4096-word memory which provides an average

instruction rate of 400,000 instructions per second and a peak instruction rate of 625,000 per second.)

Time Pulses: 14 at 160-musec intervals (4-dutnmy pulses)

Instruction Rate

Average: 275,000/sec

Peak: 450,000/sec

Instruction Word Length: 20 bits plus parity

Instruction selection: 5 bits

Index selection: 2 bits

Address selection: 13 bits

Arithmetic Speed (2-usec memory)

Add a/o subtract: 2.24 usec

Multiply: 8 usec

Circuits

Types (high-speed basic):

Dynamic: 5

Static: 2

Packages (high-speed basic): 16 types

Components:

Error detection transistors: 5,000

Operating transistors: 15,000

Logic diodes: 15,000

ELEMENTS:

Operating console, I/O typewriter, card reader, power supply module, and six logic modules. A 2-usec, 1024-word ferrite memory and a 2.5-usec, 8,000,000-bit magnetic drum memory are now used.

REFERENCE DATA:

Final report – ECPX 0022, "The Design and Use of the RTA – One for the Evaluation of Reliability Techniques," Vols. I and II. Paper – RTA Computer, by T.E. Digan.

Figure 2-56. ASC-15 Computer

ADVANCED SYSTEM CONTROLLER (ASC-15)

A modified ASC-15 is being used in the Titan II Missile; a modified version is also being utilized in the Saturn C-1 vehicle.

DESCRIPTION:

The ASC-15 computer is one of a series of advanced system controllers presently being developed as guidance and control computers. This series has evolved from nine years of activity in the development of vehicle guidance computers for the Air Force. The computers are characterized by small size, light weight, minimum power and *are* designed consistent with severe military environments. (See Figure 2 -56)

APPLICATIONS:

Guidance & Control system for Missiles and Spacecraft

FEATURES:

** Extreme flexibility
** Memory thin shell magnetic drum with air-floated heads.
** Twelve different circuit modules (except for special peripheral circuits)
**Welded encapsulated modular packaging (See Figure 2-57)

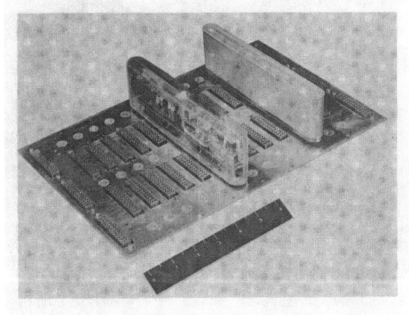

Figure 2-57 Logic Stick ASC-15

** Multiple arithmetic elements

CHARACTERISTICS:

Type - General purpose, stored program
Stored Program Drum -
     6528 Instructions
     768 Constants
     230 Line Registers

Word Length - 25 bits (including sign)

Arithmetic - Serial, fractional organization, two's complement, two independent Arithmetic Elements (adder and multiplier).

Pulse Repetition Rate - 172.8 Kc

Timing - Four clocks per bit

Command Structure - Two-address

Operation Rate -

    6400 additions per second (156 microseconds)

    543 multiplications per second (1 .87 milliseconds)

Input-Output-Capabilities –

    Inputs -

        6digital reflected binary

        48 discrete

        One pulse

    Outputs -

        3 altitude (6-bit digital to analog each)

        12 discrete

        One digital 5-bits parallel

ELEMENTS:

The ASC-15 consists of a memory (thin shell magnetic drum), an input-output unit, a control unit, and an arithmetic unit (independent arithmetic elements add-subtract, multiply, etc.)

Circuitry – Germanium Saturated Voltage Mode (diode logic)

Total Components – 7,272

Power (dc) – 150 watts

Packaging – Welded encapsulated modules

Operation Environment – Missile Environments

REFERENCE DATA:

"Advanced Systems Controllers – General Description."

MISSILEBORNE DIGITAL COMPUTER
STINGS (STELLAR INERTIAL GUIDANCE SYSTEM)

The STINGS Computer was developed under Air Force Contract No. AF04 (694)-118.

DESCRIPTION:

The Missileborne Digital Computer is a binary, fixed-point, stored program, general-purpose computer, designed for ballistic missile or space vehicle guidance.   Its memory is a random access, coincident current, ferrite array with non-destructive readout, an IBM development called RANDAM (random access, non-destructive advanced memory). The basic storage element of the RANDAM memory is a two-hole ferrite core analogous to a transfluxer, but with an improved geometry and mode of operation which reduces power consumption far below that of typical transfluxer memories.   The properties of RANDAM make it possible to read or write serially or in series-parallel. This feature allows the memory to work with a serial arithmetic unit, without a separate buffer memory.   The low-power circuits that are used in this computer permit a mechanical design without forced cooling by air or liquid. These low-power circuits were made possible by the recent evolution of epitaxial planar transistors, and the packaging techniques proposed permits a compact, miniaturized computer to be developed with minimum power density build-up.

The foremost design objective is reliability.   This implies the use of thoroughly proven hardware in a minimal computer system design providing a reliable system with a minimum of size, weight, and power.   Storage external to the memory is predominantly located in the area of the shift registers.   The failure rate in this area is minimized by using delay lines for arithmetic and memory address storage.   The delay line is an obvious choice when the transistor count is considered for the various registers.

The use of a serial memory and delay line shift registers permits an extremely simple organization.   This then reflects the design goal reliability achieved through the use of thoroughly proven hardware in an absolutely minimal computer configuration. (See Figure 2-65)

APPLICATIONS:

Ballistic Missiles, Space Vehicles

Figure 2-65 STINGS Array Vibration Model

FEATURES:

High Reliability
Low Average Power Consumption
Highly Efficient Memory
Non-Destructive Read-Out

The memory array is a matrix assemblage of MARS (Multiple Aperture Reluctance Switch) storage elements. The use of this system makes possible significant savings in system power by virtue of the nondestructive readout characteristics of the MARS device and low level standby circuitry. An advantage of this system is its contribution toward greater system reliability. The fact that information regeneration is not necessary is of major consequence. (See Figure 2-66.)

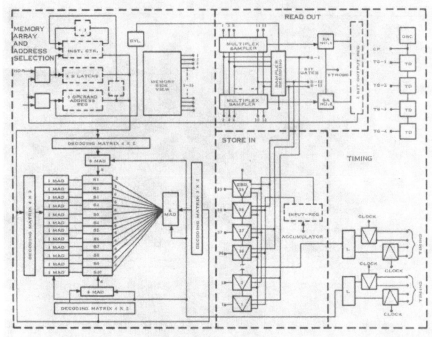

Figure 2-66 STINGS Memory Array and Address Selection

The MARS device was developed at IBM Space Guidance Center under Military Applications Research Program (now IRAD) funding during1959-1960. Concurrent with the development of the storage device, several packaging methods were evaluated. The RANDAM memory was developed primarily for use with advanced design computers utilizing microminiature circuits, but it was recognized that it would also be required with present computer systems using conventional miniature discrete components. Therefore, the RANDAM is of a modular design which is compatible for use with both present and advanced computer system design requirements.   Some of the salient characteristics on the MARS device are:

    a.   Non-destructive read-out.

    b.   Ideally suited for three-dimensional, coincident-current operation.

    c.   Half-select currents required for the read and control operations are the same magnitude.

    d.   Speed capabilities for both read and store modes are at least as fast as a three-dimensional, toroidal core memory.

e.   Device size permits matrix densities of at least 3000 bits per cubic inch.

f.   Mechanical characteristics afford relative ease of manufacture and handling.

The computer circuits operate in the saturated voltage mode using diode logic and silicon, planar, epitaxial transistors. The use of advanced components has allowed the reduction of the logic circuits to their simplest form, resulting in a substantial reduction in component count, power dissipation, size and weight.

Saturated-voltage-mode diode logic permits the most reliable circuit and over-all computer design possible at present for computers in the speed ranges required.

The computer logic design, comprising almost half of the total circuitry, utilizes only three basic types of circuits: AND, OR and inverter. These circuits are in a high state of development as the result of over one year's work by the Advanced Circuits Development Group at the SGC. (See Figure 2-67.)

CHARACTERISTICS:

Memory

4096 words (39 bits/word), random access non-destructive read-out. Division between instruction and djata storage is flexible. Typical example: 1500 data words (25 bits and sign/word) 9288 instructions (13 bits/instruction)

Arithmetic Times

Add, subtract, transfer operation - 140 microseconds, multiplication (full precision) - 420 microseconds, division (full precision) - 840 microseconds . Multiply or divide may be programmed concurrently with add, subtract, or transfer operations.

Clock Rates

500-kc arithmetic bit rate, 250-kc memory cycle rate.

Inputs

40 discretes

3 accelerometer pulse input channels

3 gimbal angle channels

one 6.4-kc (real time) counter

2 spare pulse input channels

Outputs

20 discretes

3 steering command signals gyro torquing pulses (3 gyros)

4 spare torquing pulse channels

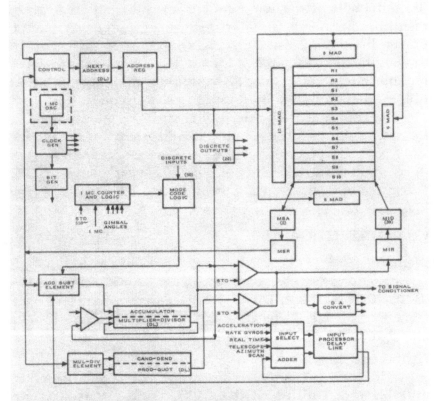

Figure 2–67 STINGS Computer – Block Diagram

Temperature Environment: 10° F to 100° F

Weight: 38 pounds

Size: 0.85 cubic feet

Power: 40 watts

PHYSICAL DESCRIPTION:

The general theory underlying the approach to the computer packaging design is one of providing environmental resistance to the most fragile subassemblies by the utilization of damped structural materials between them and the computer mounting interface, with emphasis on the most practical size and weight. The outer structure

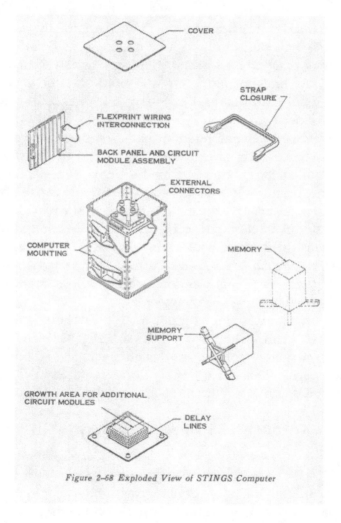

COVER

STRAP CLOSURE

FLEXPRINT WIRING INTERCONNECTION

BACK PANEL AND CIRCUIT MODULE ASSEMBLY

EXTERNAL CONNECTORS

COMPUTER MOUNTING

MEMORY

MEMORY SUPPORT

GROWTH AREA FOR ADDITIONAL CIRCUIT MODULES

DELAY LINES

Figure 2-68 Exploded View of STINGS Computer

consists primarily of laminated aluminum-visco-elastic sandwich-type material of the thinnest practical gage, fabricated by riveting, with suitable mounting appendages both externally and internally to provide mountings for subassembly components, and to the vehicle proper (See Figure 2-68).

The electronics are contained in four major subassemblies or panels, which are fastened to the interior walls of the fundamental structure. There, heat transfer conduction surfaces are under pressure-contact with the external skin of the main computer shell. The memory is located in the center of this assembly mounted on a light weight, highly damped, "X" member element. Sufficient clearance is maintained around the memory to permit 1/4-inch deflection in all directions . Delay lines are mounted on one of the covers. Interconnection to electronic panels, delay lines, memory, and external connections are made by the use of multi-layer flexible wiring, i.e., "flexprint" by Sanders Associates.

This assembly configuration has been arrived at as the optimum combination of the following general design objectives: effective utilization of the enclosed volume, simple assembly, minimum structural elements, good distortion capability with high damping capacity for vibration energy absorption, and good heat transfer characteristics to the outside environment with no requirement for internal cooling or air flow. This latter item eliminates approximately 5 lbs. required for sealed connectors, humidity resistant coatings on subassemblies, radio noise interference filters and hose connectors.

The circuit packaging concept for the computer is a simple modular structure. The basic modular size is 1 inch x 1-1/16 inches x 3/8 inch. Component density averages 30 components per cubic inch. This density is similar to the present component density in the TITAN circuit module assembly.

PRIMARY PROCESSOR & DATA STORAGE (PPDS)

STATUS

The Space Guidance Center was under contract to the Grumman Aircraft Engineering Corporation to design, build and evaluate five modes of the Primary Processor and Data Storage equipment (PPDS).

The first three models of the PPDS are prototype models to be used for functional integration and qualification testing. Two of the prototype models will be delivered to GAEC and one will be retained at IBM Owego for design evaluation. The later two models of the PPDS are flight models and both will be delivered to GAEC for incorporation into OAO spacecraft that will be placed in orbit. The first OAO was launched from the Atlantic Missile Range (Cape Canaveral) in the later half of 1963 (See Figure 2-62).

Figure 2-62 OAO (Artists Conception)

DESCRIPTION:

The PPDS equipment (See Figure 2-63) will function as the OAO spacecraft coordinator, commanding the operations to be performed, controlling the flow of information, and storing all data and instructions. The OAO viewing devices will scan celestial bodies and store celestrial measurements in the data storage equipment. The primary processor will provide positioning data for OAO's star trackers and command the sequencing of each experiment. When OAO orbits over its ground stations in the eastern United States and South America, its radios will transmit all collected observations from data storage. To safeguard against transmission errors and radio interference, data storage can be repeatedly read out. All this collected experimental data can be read out without electronic erasure. During the 10 minutes that OAO is typically in radio range of a ground station, instructions for pointing the spacecraft and sequencing experiment will be sent up and stored in its smaller command memory.

The primary processor is essentially in charge of the OAO System. It provides all system timing; verifies, decodes, and distributes both radio and stored commands; has a 256 word command error for spacecraft stabilization and control signals which govern the operation of all experiments.

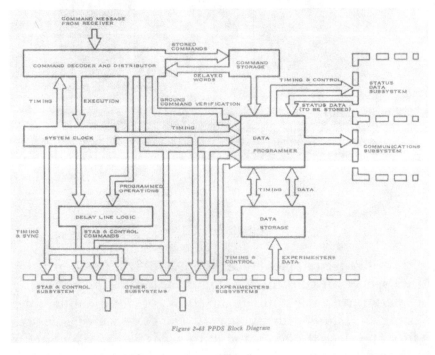

Figure 2-63 PPDS Block Diagram

The data storage is basically a large capacity ferrite memory which stores both the experimenter's data and system status data.   It can operate in a redundant (i.e. highly reliable mode giving 102,400 hits of storage or in a non-redundant mode which provides 204,800 bits of storage. Because of its non-destructive read-out capability the data storage permits multiple transmission of data to the ground station (See Figure 2-64).

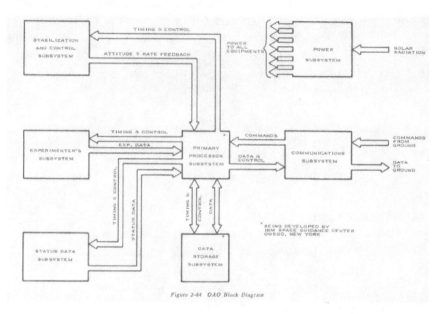

Figure 2-64 OAO Block Diagram

APPLICATIONS:
    Satellites
    Orbiting Astronomical Observatory

FEATURES:
    ** High Reliability
    ** Low Average Power Consumption
    ** Large Capacity Random Memory
    ** Highly Compact Packaging
    ** Rugged Environmental Resistance

CHARACTERISTICS:
    Design Objectives
        One year predicted reliability 0.76-0.96*
        Component Count   50,000

* *0.76 - Mode 1 Full operational capability 0.96 - Mode 2 Line-of-Sight operation*
** *Special Features*
   Note 1 - Data may be stored in successive word locations at any time. Available storage is not exhausted as time elapses between the occurrence of storable events.
   Note 2 - Memory system generates its own timing. Asynchronous operation simplifies tie-in to external equipment.
   Note 3 - Nondestructive read-out permits multiple read-out of the entire memory while satellite is in range of receiving station; thereby, improving reliability of transmitted date under poor signal to noise conditions.

   Average Power, 40 watts
   Weight, 152 lbs.
   Volume, 5 cu. Ft.
   Vibration, 7.5G to 3000 CPS

Data Storage Characteristics
   Storage capacity:
         102,400 bits (64 x 64 x 25) redundant
         204,800 bits non-redundant
   Storage efficiency (Note 1 above)**100 percent
   Timing (Note 2 above)** Asynchronous
   Store Rate: 0 to 2,500,000 bits per sec (For 25 binary bit parallel input)
   Read rate (Note 3 above) ** 0 to 100,000 bits per sec (serial output)

Packaging
   Redundancy Triple modular and quadruple component redundancy
   Schedule 9-12 mos. for prototype production model

ELEMENTS:

   The PPDS equipment is divided into 12 functions. These functions are packaged in two units — Command Processor and Data Storage, and Control Processor and Command

Storage. Functions for the Command Processor and Data Storage are:

** Receive and verify
** Data programmer
** Command decoder and distributor
** System clock
** Data storage

Functions for the Control Processor and Command Storage are:

** Attitude delay line logic
** Gimbal delay line logic
** Experiment register and decoder
** Control matrix
** Programmer startracker signal controller
** Power converter
** Command storage

The internal packaging design for these units centers around pluggable modules and panel assemblies. Exceptions to this design are the unique electrical component or assemblies — memory arrays, delay lines and power converter. The pluggable modules consist of two assemblies — the housing or cell with its contacts and the welded circuit assembly.

The cell panels provide the mounting, support and electrical interconnections for the pluggable modules. A unit contains a maximum of 24 cell panels.

REFERENCE DATA:

Primary Processor and Data Storage Phase I Design Report, IBM File No. 61-807-14, May 15, 1961

Orbiting Astonomical Observatory to Grumman Aircraft Engineering Corp., IBM File No. 61-807-6A, April 15, 1961.

OAO Computer Functional Capabilities by E.S. Flanders, February 28, 1961

The Orbiting Astronomical Observatory-A Description, IBM File No. 61-503-1 April 18, 1961.

INFORMER SYSTEM

Developed under U.S. Army Signal Corps Contract.

APPLICATIONS:

Informer Processor:
    Reconnaissance operating
    Fire control and planning
    Ground guidance systems
    Missile checkout system
    Topographic data processing
Retriever Processor:
    Battlefield surveillance
    Logistics and administration
    Technical intelligence
    Bomb damage assessment
    Information retrieval
Communication Processor:
    Message switching
    Line switching
    Tactical communications
    Integrated data processing

DESCRIPTION:

The Informer is a solid-state information and retrieval system capable of storing large amounts of information and, upon request, of retrieving desired information through a search function. At the heart of this system is a small, mobile, general -purpose computer called the Informer Processor. This stored program computer can be used in a variety of military fields or shipboard applications. The Informer is unique in its utilization of pulsed magnetic circuits that employ tape cores and silicon transistors.

The Informer is constructed using the modular concept to allow for future expansion of capability of the system.

SYSTEM CONFIGURATIONS:

Informer Processor
This unit is the basic building block and central

processing unit for all Informer configurations. It is composted of an arithmetic and logical unit, a magnetic core memory, and an input-output (I/O) converter unit (See Figure 2-17).

Retriever Processor

The addition of mass storage units and high-speed search units to the Informer Processor provides information storage and retrieval capability which achieve the Retriever/Processor configuration.

Figure 2-15 Informer System (Left View)

Communication Processor

By the addition of a real-time clock, a multi-channel

communication converter, available on a rental basis. The following list is representative of this equipment:

IBM Advanced Solid State Tape Drive
1402 High-Speed Card Reader-Punch
1403 High-Speed Printer
1407 Console Inquiry Station
1405 Magnetic Disk File

CHARACTERISTICS:

Core Storage Unit

The core storage unit is a high-speed, random-access device for storage of program and data.

4096-word storage capacity
38-bit word
1 to 7 units can be used
8-microsecond cycle time
Random-access, in less than 3 microseconds after demand
Nonvolatile; does not lose information when power is off

Central Processor

The central processor performs the function of a stored-program data processor.

Binary-parallel operation
Fixed point operation
30,000 instructions per second for the average program
55 instructions, standard U.S. Army Fieldata instruction set
4 index registers, expandable to 7
Generally, single address instructions but with several double address instructions

I/O Converter

The I/O Converter executes I/O operations while the computer executes the arithmetic operations.

Simultaneous operation of 1 to 4 devices using 1 to 4 I/O converters, respectively 8-bit (6 data, 1 control, and 1 parity), parallel, character input/output with devices

Controls up to 63 I/O devices, regardless of the number of I/O converters I/O devices controlled:

Magnetic tape punch
Paper tape punch
Paper tape reader
I/O typewriter
Disk file
Card reader/punch
High-speed printer

REFERENCE DATA:

Informer General Information Manual, May 1962
Informer Preliminary Reference Manual Informer Brochure No. 550-005

ACOUSTIC TRACKING SYSTEM

Work performed under DTMB Navy Contract N/600 (167)-56854(x) at Owego.

DESCRIPTION:

A deep-water acoustic range is being developed by the David Taylor Model Basin for the Bureau of Ships in a program concerned with minimizing the sound radiated by operational submarines. In order to secure accurate acoustic signatures of the submarine, range and bearing information, obtained by the Acoustic Tracking System, is employed to monitor and control the submarines movement.

For an accurate evaluation of the acoustic signatures of the submarine, a fully operational acoustic tracking system, employing automated techniques, would have to accomplish the following:

** Track a submerged submarine, traveling at varying speeds (possibly a nuclear powered submarine), relative to a drifting hydrophone array as the submarine attempts to maintain a constant velocity, constant depth, and a predetermined course. Accuracies of + 3% in range and +_2% in bearing, are required.

** Communicate to the submarine appropriate up-dated information to aid in the navigation of the submarine along the prescribed course.

** Provide a visual display of the submarine track.

** Provide analog signals, proportional to the submarine range and bearing, suitable for recording on magnetic tape for later analysis.

FEATURES:

System components include a projector-receiver mounted on the submarine to be tracked, a projector and a receiver attached to a hydro-phone array, and a receiver attached to a second array. The two arrays are approximately 100 yards apart, and the transducers are attached at a depth of about 25 feet. Under the control of a highly accurate crystal, the projector-receiver on the submarine emits a ping that is received by the transducers on the arrays. With a knowledge of the velocity of sound in the water (obtained with a 0.1% accuracy from bathythermorgam readings), depth of the submarine (provided by DTMB), depth of the receiver (controlled) and the base line between the receivers (determined by pinging from the projector on one array to the receiver on the other) the range and bearing are computed from the elapsed time interval between the transmission of the submarine and reception by the transducer on the array. (See Figure 2-49.)

The system aboard the ship includes a digital computer which performs the required computations and control functions as well as determining whether the projector-receiver attached to the hydrophone array is transmitting or receiving. Various modes of operation can be auto-matically programmed and selected by a mode switch. The computer provides outputs through digital-to-analog converters to an X-Y plotter for the submarine track and dc voltages proportional to range and bearing. Binary data from the computer is amplified, transmitted to a com-munications projector, and then transmitted to the underwater telephone system (UQC) receiver on the submarine.

The submarine processor receives the binary data from the communications receiver. Decoding equipment interprets this data and obtains from it the submarine range bearing and track deviation. This data is used as an aid in navigating the submarine to the desired track past the hydrophone array.

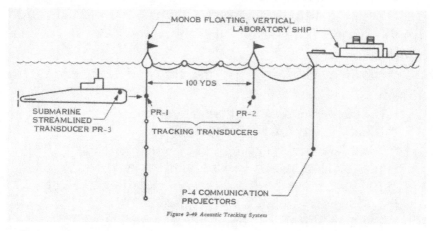

Figure 3-49 Acoustic Tracking System

ELEMENTS:

Input/Output and Conversion Equipment

** Electric Typewriter - used to enter data and to print selected results, thereby providing a permanent record.

** Paper Tape Reader - used to enter blocks of data or program material into the computer.

** Plotting Table - an X-Y plotter used to record submarine position.

** Control Outputs - control outputs are provided for discrete bit equipment control.

** Analog Output Registers - two to the X-Y plotter, and two to magnetic tape.

** Communications - an output is provided for transmission of data from the processor to the submarine.

X-Y Plotter

** 11 inches x 17 inches graph paper.
** Vacuum hold-down system.
** Static accuracy - 0.075 percent of full scale.

** Dynamic accuracy - 0.1 percent of full scale at plotting speeds of 10 inches per second.

Submarine Processor

** Gate enables transducer to be used as either a receiver or a transmitter

** Pulse modulated information is processed through dippers and transferred to shift register as digital information.

** Parity checking is used to detect errors in transmission.

** In-line decimal visual display equipment.

** Binary Control Decimal format.

** 7 bit data words (1 Flag Bit, 5 Data Bits, 1 Parity Bit).

** 10-7 Bit Data Words for Complete message.

** 10 bits for data portion of communication message.

** 1 Parity Bit per message segment.

** Total of 14 bits per communication message.

** Bit rate - Variable from 1.64 milliseconds to 26.2.

STORED PROGRAM DDA

Built under Contract No. AF 33(600)-31315. Completed Nov. 1959. Out of Production.

DESCRIPTION:

A high-speed digital differential analyzer (HS/DDA), suitable for real-time simulation and air-to-surface missile guidance applications was developed. This is a small light-weight computer having a high, variable solution rate. This design features the practical use of welded encapsulated modules and the logical combination of microalloy drift transistor circuits with a sonic delay line memory

The selection of these germanium circuits, operating at a PRF of four-megacycles and the one-megacycle magnetostrictive delay line memory was based on reliability design goals, desired computation speeds and the demand for operating under adverse environmental conditions including nuclear radiation burst contamination.

The design philosophy of the DDA was logical optimization wherein either the ternary and/or binary transfer mode is employed with a rectangular integration scheme. This approach yields the necessary accuracy for the missile system studied in view of the following considerations:

** A high iteration rate is provided by the four-megacycle circuitry and one-megacycle delay lines so that additional accuracy can be realized through (1) program-scaling while staying within solution rate requirements, and (2) the use of the variable solution rate control mode.

** Variable word lengths are provided so that the programmer has great scaling flexibility and "dead space" can be completely eliminated on the delay lines; this ultimately leads to improved accuracy.

APPLICATIONS:

The high-speed digital differential analyzer can be used to solve characteristic cruise and ballistic missile guidance and control equations.

CHARACTERISTICS:

Number System - Fixed Fractional Binary
Data Word Size - Variable
Data Word Format - Two logic bits, others for magnitude and sign
Program Word Size - Variable
Program Word Format - Two timing bits, others for transfer direction.
Clock Rate - 4 megacycles Bit Rate - 1 megacycle
Iteration Rate - 623 Solutions/sec, (nominal)
Capacity - 73 Integrators (with 22 bit words)
Operation - Serial Integration - Rectangular Timing - 4 megacycle oscillator
Increment Transfer - Binary (Option for ternary)
Volume - 0.73 cubic feet
Weight - 30.5 lbs.

Power Requirements - 29.1 watts
Data Storage - Magnetostriotive Delay Line
Program Storage - Magnetostrief ive Delay Line
Logic - Pulse, level
Transistor Type-Germanium, Micro-alloy drift
Circuit Packaging - Welded Encapsulated Modules
Transistors - 309
Diodes - 886
Total Components - 3114
Supply Voltages - +9.5 ±5%; -9.5 ±5%; -3.5 ±3%

FEATURES:

Variable Integrator Length - By eliminating dead time the integrator capacity of the DDA can be increased or a portion of the storage can be eliminated.   Elimination of some storage requirements by using this technique results in an increased iteration rate.   Depending on how it is used, the feature can increase accuracy, speed, or storage capacity, since they are all interdependent.

Automatic Programming - A 704 program has been written to automatically code the program of the DDA. Basically the program will consist of a transformation of a DDA integrator configuration into coded cards magnetic tape core memory which can be loaded via the control unit,   This automatic coding will substantially reduce the programming time for large configurations of integrators.

Circuits

The design of the high-speed digital differential analyzer circuits was based on the following considerations:
100,000 hour reliability four-mega cycle operating speeds small signal transmission noise rejection circuit standardization component and power voltage standardization economy.

Packaging

The WEM was specifically developed for maximum reliability under severe environmental stresses.   This technique was a natural out-growth of the effort to obtain highly reliable connections.   A great deal of investigation into the metallurgical joining of materials by welding was necessary in the refinement of this package. A similar effort was expended in the investigation of encapsulating agents that would be compatible with this packaging concept.

A NOTE ON SOURCES

I have used a great deal of documentation available to the public. Much is previously known to technical scholars and to any reader who has kept abreast of both the computer world and our nation's defense posture. However, many details of IBM's military contracts have never before been shown or been published. The nineteen books written over the past twenty-five years about the IBM Corporation by both insiders and others have, for the most part, generally ignored the Federal Systems Division's role and its brilliant workforce. For, after all, FSD was IBM, and IBM was FSD. IBM itself, even in today's online historical archives, fails time and time again to fully distinguish Federal Systems contributions.

IBM's archives, the Internet's online Wikipedia, as well as other search engines, contain much history on IBM, NASA and the United States military. The basic narrative is taken from many of these sources, including corporate annual reports of both Loral and Lockheed-Martin. While I interviewed many former IBMers, most were ignorant of the role played by this government division.

Technical specifications on FSD's projects and products were developed from recent interviews and from thirty-five-year-old personal papers. I am indebted to all those sources. The selected bibliography which follows comprises some of the most important of these sources.

All photographs and technical drawings shown were produced almost fifty years ago and can be credited only to unknown FSD/IBM in-house photographers and/or government agency sources.

-- W.L.R.

BIBLIOGRAPHY

BOOKS

BAMFORD, James. *Body of Secrets*, Doubleday, 2001

BELDEN, Thomas Graham. BELDEN, Marva Robins. *The Lengthening Shadow*, Little, Brown & Company, 1962

BAUER, Roy A. COLLAR, Emilio. TANG, Victor. *Silverlake, Project, Transformation At IBM*, Oxford U. Press, 1992

BLACK, Edwin. *IBM And The Holocaust*, Crown Publishers, N.Y. 2001

CARROLL, Paul. *Big Blues, The Unmaking of IBM*, Crown Publishing, 1993

CHPOSKY, James. LEONSIS, Ted, *Blue Magic*, 1988

DELAMARTER, Richard Thomas. Big Blue, *IBM's Use & Abuse of Power*. McMillian, 1986

FERGUSON, Charles H., MORRIS, Charles R. *Computer Wars*, Times Books, Random House, 1994

FOY, Nancy., *The Sun Never Sets on IBM*, William Morrow & Company Inc, 1975

GARR, Doug. *IBM Redux*, Harper Business, 1999

GERSTNER, Jr., Louis V. *Who Says Elephants Can't Dance*, Harper Business, 2002

HUMPHREY, Watts S. *Managing For Innovation*, Prentice Hall, 1987

KILLEN, Michael. *IBM The Making Of The Common View*, Harcourt Brace & Jovaorvich, 1988

MANEY, Kevin. *The Maverick and His Machine*, John Wiley & Sons, 3003

MCKENNA, Regis. *Who's Afraid of Big Blue?*, Addison-Wesley
Publishing Inc. 1989.

MOBLEY, Lou. McKEOWN, Kate. *Beyond IBM*, McGraw Hill, 1986

PETRE, Peter. Father & Son. Bantam Books, 1990

ROGERS, F.G. "Buck". *The IBM Way*, Harper & Row, N.Y. 1986   SLATER, Robert. *Saving Big Blue*, McGraw Hill,

1999

SOBEL, Robert. *IBM Colossus In Transition*, 1981

SOBEL, Robert. *IBM vs. Japan*, Stein & Day, 1986

TAUBMAN, Philip. *Secret Empire*, Simon & Schuster, 2003

TEDLOW, Richard S. *The Watson Dynasty*, Harper Business 2003

WITHINGTON, Frederic G. *The Use of Computers in Business Organizations*, Addison-Wesley Publishing, 1966

ARCHIVES AND LIBRARIES

Barnes & Noble Booksellers, Reference

Roswell Public Library, Roswell, Georgia

Sandy Spring Library, Sandy Spring, Georgia

IBM Archives - Internet

NASA Space History - Internet

Earthlink, IBM Navy Submarines - Internet

Defense Acquisition History Project - Internet

Wikipedia Internet Encyclopedia, FAA, DOD, DIA,IBM

IBM Journal Of Research & Development - Internet

ARMY, Redstone Military History - Internet

DACS Military History Modern Programming Practices - Internet

USAF Air Power, Maxwell AFB - Internet

US NAVY, Military Library, Naval Intelligence Ops - Internet

GLOBAL SECURITY, WMD, NMCC On-Line Internet J-2 Joint Staff Intelligence, Internet

R&D Expenditures, Statistics All Countries Org. - Internet

NSF Government Patterns of R&D Resources, Internet

Lockheed Martin Systems Integration Owego – Internet

NEWSPAPERS, MAGAZINES AND PERIODICALS

The Washington Post

The New York Times

The Atlanta Constitution

Aviation Week & Space Technology, 1950-2007

Washington Technology

Newsweek, 1987-2003

Business Week, 1990-1995

Forbes, 1990-2004

Fortune, 1991-1993

Police Department-City of New York, Office Of Management Analysis & Planning

Multinational MONITOR, Internet

The Wall Street Journal U.S. News & World Report, 1987-2004   Lockheed Martin Annual Report 1966

Loral Annual Reports 1994, 1995

DATA Publications, *R&D Proposal Preparation Guide* 1962

DATA Publications, *Winning R&D Proposals Through Critiques*, 1963

IBM Quarter Century Club, Atlanta Chapter Internet

SPACE DATA, TRW Systems, 1965

INTERVIEWS & ORAL HISTORIES

Fred H. Lippucci, 2007

George Liptak, 2006, 2007

Robert Campenni, 2007

Marilyn Wedig, 2007

Jerry Golem, 2006, 2007

Robert J. Blum, 2007

James Wilson, OWL Group, 2007

Donald Norris, 2006, 2007

The OWL Group, Ex-IBMers

Alvin Ginsburg, OWL Group

Ed Glass, OWL Group, 2007

Mike McGuirt, 2007

Richard Massey, 2007

Steve Jackson, Loral Space & Communications, 2007

INDEX

A

William Louis Robinson's
Scam Stars of the Airwaves
Coming in 2009 from ThomasMax Publishing!

They thought they were getting in on the ground floor,
but in reality they were being taken

Scam Stars of the Airways is the story of a massive, larcenous, nationwide scam that shook down the American public for more than $60 million in the early 1990s. Ponzi's "Something-for-Nothing" Pyramid Scheme of the 1920s pales in comparison to this brilliantly conceived and flawlessly executed national rip-off involving the nation's airwave-licensing rights. The tale features starry-eyed investors, a government agency mired in bureaucratic controversy, and the smoothest pitchmen to come along since Elmer Gantry!

This national "shell game" ensnarled ordinary folks, lawyers and stockbrokers, bankers and fiduciaries – and a judge. Class-action suits were filed against the boiler-room operators as they moved on with states' cease-and-desist orders flapping in the breeze.

Federal and state regulators, for the most part, missed the real behind-the-scenes story . . . and most of the true culprits who masterminded this colossal TV wireless-venture swindle.

Robinson's tale is fascinating and humorous, and one that every investor would do well to read.

Printed in the United States
109863LV00004B/144/P